W9-BXX-427

Lewis Carroll in Numberland

No carte has yet been done of me,
   that does real justice to my *smile*;
and so I hardly like, you see,
   to send you one — however, I'll
consider if I will or not —
   meanwhile, I send a little thing
to give you an idea of what
   I look like when I'm lecturing.
The merest sketch, you will allow —
   yet still I think there's something grand
In the expression of the brow
   and in the action of the hand.

Lewis Carroll in Numberland

His Fantastical Mathematical Logical Life

An Agony in Eight Fits

ROBIN WILSON

W. W. Norton & Company
New York · London

First published 2008 by Allen Lane, an imprint of Penguin Books, London

Copyright © 2008 by Robin Wilson
First American Edition 2008

All rights reserved
Printed in the United States of America
First published as a Norton paperback 2010

For information about permission to reproduce selections from this book, write to
Permissions, W. W. Norton & Company, Inc., 500 Fifth Avenue,
New York, NY 10110

For information about special discounts for bulk purchases, please contact
W. W. Norton Special Sales at specialsales@wwnorton.com or 800-233-4830

Manufacturing by Courier Westford
Production manager: Devon Zahn

Library of Congress Cataloging-in-Publication Data

Wilson, Robin J.
Lewis Carroll in numberland : his fantastical mathematical logical
life : an agony in eight fits / Robin Wilson.— 1st American ed.
p. cm.
Includes bibliographical references and index.
ISBN 978-0-393-06027-0 (hardcover)
1. Mathematicians—Biography. 2. Carroll, Lewis,
1832–1898—Knowledge—Mathematics. 3. Mathematics. I. Title.
QA29.C353W55 2008
510.92—dc22
[B]
2008037342

ISBN 978-0-393-30452-7 pbk.

W. W. Norton & Company, Inc.
500 Fifth Avenue, New York, N.Y. 10110
www.wwnorton.com

W. W. Norton & Company Ltd.
Castle House, 75/76 Wells Street, London W1T 3QT

1 2 3 4 5 6 7 8 9 0

Contents

Preface

Charles Dodgson is best known for his *Alice* books, *Alice's Adventures in Wonderland* and *Through the Looking-Glass*, written under his pen-name of Lewis Carroll. These books have delighted children and adults for generations and have never been out of print.

If Dodgson had not written the *Alice* books, he would be remembered mainly as a pioneering photographer, one of the first to consider photography as an art rather than as simply a means of recording images. In particular, his imaginatively posed photographs of children are a delight, and his hundreds of photographs of friends and celebrities provide us with much insight into the Victorian world around him.

If Dodgson had not written the *Alice* books or been a photographer, he might be remembered as a mathematician, the career he pursued as a lecturer at Christ Church, the largest college of Oxford University. But what mathematics did he do? How good a mathematician was he, and how influential was his work?

In this book, written for a general readership, I try to answer these questions. In particular, I describe his work in geometry, algebra, logic and the mathematics of voting, in the context of his other activities, and on the lighter side I present some of the puzzles and paradoxes with which he delighted in entertaining his child-friends and contemporaries. Much work has been done on his contributions to all these areas; my aim here is to make this material accessible to a wider readership.

I am grateful to many people who have helped me with the preparation of this book. When I first became interested in Charles Dodgson I received much help and encouragement from Francine Abeles, who has edited the mathematical and political pamphlets of Charles Dodgson (see the Notes and References) and who has sent me much useful material on algebra, voting, ciphers, logic, and other topics.

Later I was privileged to get to know Edward Wakeling, who edited Dodgson's Oxford pamphlets, produced two popular books of Dodgson's puzzles, and has undertaken the Herculean task of editing the Dodgson diaries in ten volumes — a work of great scholarship and an invaluable source for anyone interested in the facts, rather than the myths, of Dodgson's life. I am particularly grateful to him for allowing me access to his magnificent archive of Carrolliana and for freely giving his time to introduce me to much material with which I was unfamiliar, for providing me with a great deal of useful information, and for correcting many errors in my manuscript.

Finally, I wish to thank Mark Richards, President of the Lewis Carroll Society, and Amirouche Moktefi for their help with several sections of this book. I should also like to thank John Woodruff for his careful editing.

Robin Wilson
Oxford, April 2008

Chronology of Events

This is not a full chronology of Charles Dodgson's life, but contains the milestones and his most important mathematical (and other) publications. Several titles are abbreviated.

1832 27 January: born at Daresbury, Cheshire

1843 Moves to Croft Rectory, Yorkshire

1844 Attends Richmond Grammar School

1846 Attends Rugby School

1849 Returns to Croft Rectory

1850 Matriculates at Oxford University

1851 Takes up residence at Christ Church, Oxford
      Mother dies

1852 Elected a 'Student' of Christ Church

1854 Long vacation at Whitby studying with Bartholomew Price
      First Class in Mathematics in his Finals Examinations
      Receives Bachelor of Arts degree

1855 Begins teaching at Christ Church
      Henry Liddell appointed Dean of Christ Church
      Elected Mathematical Lecturer at Christ Church

1856 Adopts the pseudonym Lewis Carroll
      Begins hobby of photography

1857 Receives Master of Arts degree
      'Hiawatha's Photographing'

1860 *A Syllabus of Plane Algebraic Geometry*
      Notes on the First Two Books of Euclid

1861 *Notes on the First Part of Algebra*
      The Formulae of Plane Trigonometry
      Ordained Deacon by Bishop Wilberforce

1862 Boat trip to Godstow with the Liddell sisters

1863 *The Enunciations of Euclid I, II*

1864 *A Guide to the Mathematical Student*
Completes the manuscript of *Alice's Adventures Under Ground*

1865 *The Dynamics of a Parti-cle, with an Excursus on the
New Method of Evaluation as Applied to π*
Alice's Adventures in Wonderland

1866 'Condensation of Determinants'

1867 *An Elementary Treatise on Determinants*
Tour of the Continent with Dr Liddon

1868 Father dies, and the Dodgson family moves to Guildford
The Fifth Book of Euclid Treated Algebraically
Algebraical Formulae for Responsions
Moves into a new suite of rooms in Tom Quad

1869 *Phantasmagoria and other Poems*

1870 *Algebraical Formulae and Rules*
Arithmetical Formulae and Rules

1871 *Through the Looking-Glass, and What Alice Found There*

1872 *Symbols, &c., to be used in Euclid, Books I and II*
Number of Propositions in Euclid

1873 *The Enunciations of Euclid I–VI*
*A Discussion of the Various Methods of Procedure in
Conducting Elections*

1874 *Suggestions as to the Best Method of Taking Votes*
Examples in Arithmetic

1876 *The Hunting of the Snark*
A Method of Taking Votes on More than Two Issues

1877 First summer holiday in Eastbourne

1879 *Euclid and his Modern Rivals*

1880 Proposes reduction in salary

1881 Resigns Mathematical Lectureship

1882 *Euclid, Books I, II* (earlier unpublished edition, 1875)
Becomes Curator of Christ Church Common Room

1883 *Lawn Tennis Tournaments*

1884 *The Principles of Parliamentary Representation*

1885 *A Tangled Tale*

1886 *Alice's Adventures Under Ground* (facsimile edition)

1887 *The Game of Logic* (earlier private edition, 1886)
To Find the Day of the Week for Any Given Date

1888 *Curiosa Mathematica*, Part I: *A New Theory of Parallels*
Memoria Technica

1889 *Sylvie and Bruno*

1890 *The Nursery "Alice"*

1891 Henry Liddell resigns as Dean of Christ Church

1892 Resigns as Curator of Christ Church Common Room

1893 *Curiosa Mathematica*, Part II: *Pillow-Problems*
Sylvie and Bruno Concluded

1894 *A Logical Paradox*

1895 'What the Tortoise Said to Achilles'

1896 *Symbolic Logic,* Part I: *Elementary*

1897 *Brief Method of Dividing a Given Number by 9 or 11*
Abridged Long Division

1898 14 January: dies in Guildford

The Mock Turtle tells Alice his sad story

Introduction
From Gryphons to Gravity

"Begin at the beginning," the King said, very gravely,
"and go on till you come to the end: then stop."

As you might expect from a lecturer in mathematics, Lewis Carroll's books for children are brimming with mathematical allusions — arithmetical, geometrical, logical and mechanical. This is the world of mock turtles and maps, gryphons and gravity, Humpty Dumpty and handkerchiefs — recast here in dramatic form in eight scenes.

Scene 1: The Mock Turtle's Education

In *Alice's Adventures in Wonderland* (1865), Alice is introduced to the Gryphon, who leads her to a rocky seashore. There they encounter the Mock Turtle, who looks at them with large eyes full of tears.

Mock Turtle: Once I was a real turtle.

Gryphon: Hjckrrh!

Mock Turtle: When we were little, we went to school in the sea. The master was an old turtle — we used to call him Tortoise —

Alice: Why did you call him Tortoise, if he wasn't one?

Mock Turtle: We called him Tortoise because he taught us. Really you are very dull!

Gryphon: You ought to be ashamed of yourself for asking such a simple question.

Mock Turtle: Yes, we went to school in the sea. I only took the regular course.

Alice: What was that?

Mock Turtle: Reeling and Writhing, of course, to begin with; and then the different branches of Arithmetic — Ambition, Distraction, Uglification and Derision.

Alice: I never heard of 'Uglification'. What is it?

1

Gryphon: Never heard of uglifying! You know what to beautify is, I suppose?

Alice: Yes: it means — to — make — anything — prettier.

Gryphon: Well, then, if you don't know what to uglify is, you *are* a simpleton.

Alice: And how many hours a day did you do lessons?

Mock Turtle: Ten hours the first day, nine hours the next, and so on.

Alice: What a curious plan!

Gryphon: That's the reason they're called lessons — because they lessen from day to day.

Alice: Then the eleventh day must have been a holiday.

Mock Turtle: Of course it was.

Alice: And how did you manage on the twelfth?

Gryphon: That's enough about lessons.

Scene 2: Humpty Dumpty's Cravat

In Lewis Carroll's second *Alice* book, *Through the Looking-Glass* (1871), Alice encounters the argumentative Humpty Dumpty, a stickler for the meaning of words, for whom a simple arithmetical calculation proves to be rather a challenge.

Humpty: Tell me your name and your business.

Alice: My *name* is Alice, but —

Humpty: It's a stupid name enough! What does it mean?

Alice: *Must* a name mean something?

Humpty: Of course it must: *my* name means the shape I am — and a good handsome shape it is, too. With a name like yours, you might be any shape, almost. How old did you say you were?

Alice: Seven years and six months.

Humpty: Wrong! You never said a word like it!

Alice: I thought you meant 'How old *are* you?'

Humpty: If I'd meant that, I'd have said it.

Alice: (*after a pause*) What a beautiful belt you've got on! At least, a beautiful cravat — no, a belt, I mean — I beg your pardon!

Humpty: It's a cravat, child, and a beautiful one, as you say. It's a present from the White King and Queen. They gave it me — for an un-birthday present.

Humpty Dumpty sat on a wall

Alice: I beg your pardon?

Humpty: I'm not offended.

Alice: I mean, what *is* an un-birthday present?

Humpty: A present given when it isn't your birthday, of course.

Alice: I like birthday presents best.

Humpty: You don't know what you're talking about! How many days are there in a year?

Alice: Three hundred and sixty-five.

Humpty: And how many birthdays have you?

Alice: One.

Humpty: And if you take one from three hundred and sixty-five, what remains?

Alice: Three hundred and sixty-four, of course.

Humpty: I'd rather see that done on paper.

Alice: Three hundred and sixty-five . . .   365
                    minus one . . .   – 1
      is three hundred and sixty-four.   364

Humpty: That seems to be done right —

Alice: You're holding it upside down!

Humpty: To be sure I was! I thought it looked a little queer. As I was saying, that *seems* to be done right — though I haven't time to look over it thoroughly right now — and that shows that there are three hundred and sixty-four days when you might get un-birthday presents —

Alice: Certainly.

Humpty: And only *one* for birthday presents, you know. There's glory for you!

Alice: I don't know what you mean by 'glory'.

Humpty: Of course you don't — till I tell you. I meant 'there's a nice knock-down argument for you!'

Alice: But 'glory' doesn't mean 'a nice knock-down argument'.

Humpty: When *I* use a word, it means just what I choose it to mean — neither more nor less.

Scene 3: Alice's Examination

When Alice finally reaches the Eighth Square on the looking-glass chessboard, she expects to become Queen — but first she must be interrogated by the Red Queen and the White Queen.

Red Queen: You ca'n't be a queen, you know, till you've passed the proper examination. And the sooner we begin it, the better.

White Queen: Can you do Addition? What's one and one and one and one and one and one and one and one and one and one?

Alice is examined by the White Queen and the Red Queen

Alice: I don't know. I lost count.

Red Queen: She ca'n't do Addition. Can you do Subtraction? Take nine from eight.

Alice: Nine from eight I ca'n't, you know: but —

White Queen: She ca'n't do Subtraction. Can you do Division? Divide a loaf by a knife. What's the answer to *that*?

Alice: I suppose —

Red Queen: Bread-and-butter, of course. Try another Subtraction sum. Take a bone from a dog: what remains?

Alice: The bone wouldn't remain, of course, if I took it — and the dog wouldn't remain: it would come to bite me — and I'm sure *I* shouldn't remain!

Red Queen: Then you think nothing would remain?

Alice: I think that's the answer.

Red Queen: Wrong, as usual. The dog's temper would remain.

Alice: But I don't see how —

Red Queen: Why, look here! The dog would lose its temper, wouldn't it?

Alice: Perhaps it would.

Red Queen: Then if the dog went away, its temper would remain!

Both Queens: She ca'n't do sums a *bit*!

Scene 4: What's in a Name?

Logical and philosophical absurdities permeate the *Alice* books — such as the Cheshire Cat's celebrated grin in *Alice's Adventures in Wonderland*:

> "All right," said the Cat; and this time it vanished quite slowly, beginning with the end of the tail, and ending with the grin, which remained some time after the rest of it had gone.
>
> "Well! I've often seen a cat without a grin," thought Alice; "but a grin without a cat! It's the most curious thing I ever saw in all my life!"

In *Through the Looking-Glass*, the White Queen challenges Alice about the nature of belief and the impossible:

> White Queen: Let's consider your age to begin with — how old are you?
>
> Alice: I'm seven and a half exactly.

White Queen: You needn't say 'exactually': I can believe it without that. Now I'll give *you* something to believe. I'm just one hundred and one, five months and a day.

Alice: I ca'n't believe *that*!

White Queen: Ca'n't you? Try again: draw a long breath, and shut your eyes.

Alice: There's no use trying; one *ca'n't* believe impossible things.

White Queen: I daresay you haven't had much practice. When I was your age, I always did it for half-an-hour a day. Why, sometimes I've believed as many as six impossible things before breakfast.

After her meeting with Humpty Dumpty, Alice comes across the White King, who is busily trying to protect his crown from the Lion and the Unicorn:

White King: I've sent them all! Did you happen to meet any soldiers, my dear, as you came through the wood?

Alice: Yes, I did: several thousand, I should think.

White King: Four thousand two hundred and seven, that's the exact number. I couldn't send all the horses, you know, because two of them are wanted in the game. And I haven't sent the two

Messengers, either. They're both gone to the town. Just look along the road, and tell me if you can see either of them.

Alice: I see nobody on the road.

White King: I only wish *I* had such eyes. To be able to see Nobody! And at that distance too! Why, it's as much as *I* can do to see real people, by this light!

Once Haigha, the Messenger, arrives, he is quizzed in a similar vein:

White King: Who did you pass on the road?

Haigha: Nobody.

White King: Quite right: this young lady saw him too. So of course Nobody walks slower than you.

Alice meets the White Knight

Haigha: I do my best. I'm sure nobody walks much faster than I do!

White King: He ca'n't do that, or else he'd have been here first.

Alice's next encounter is with the White Knight, and she becomes involved in a discussion about his various inventions. The conversation then turns to the naming of things:

White Knight: You are sad: let me sing you a song to comfort you.

Alice: Is it very long?

White Knight: It's long, but it's very, *very* beautiful. Everybody that hears me sing it — either it brings the *tears* into their eyes, or else —

Alice: Or else what?

White Knight: Or else it doesn't, you know. The name of the song is called '*Haddocks' Eyes*'.

Alice: Oh, that's the name of the song, is it?

White Knight: No, you don't understand. That's what the name is *called*. The name really *is* '*The Aged Aged Man*'.

Alice: Then I ought to have said 'That's what the *song* is called'?

White Knight: No, you oughtn't: that's quite another thing! The *song* is called '*Ways and Means*': but that's only what it's *called*, you know.

Alice: Well, what *is* the song, then?

White Knight: I was coming to that. The song really *is* '*A-sitting On A Gate*': and the tune's my own invention.

Scene 5: The Beaver's Lesson

Arithmetical ideas also feature in Lewis Carroll's other books for children. In Fit the Fifth of *The Hunting of the Snark* (1876), 'An Agony in Eight Fits', the scream of the dreaded Jubjub bird is heard:

> Then a scream, shrill and high, rent the shuddering sky,
>   And they knew that some danger was near:
> The Beaver turned pale to the tip of his tail,
>   And even the Butcher felt queer . . .
>
> " 'Tis the voice of the Jubjub!" he suddenly cried.
>   (This man, that they used to call "Dunce".)

"As the Bellman would tell you," he added with pride,
  "I have uttered that sentiment once."

" 'Tis the note of the Jubjub! Keep count, I entreat;
  You will find I have told it you twice.
'Tis the song of the Jubjub! The proof is complete,
  If only I've stated it thrice."

The Beaver had counted with scrupulous care,
  Attending to every word:
But it fairly lost heart, and outgrabe in despair,
  When the third repetition occurred.

It felt that, in spite of all possible pains,
  It had somehow contrived to lose count,
And the only thing now was to rack its poor brains
  By reckoning up the amount.

"Two added to one — if that could but be done,"
  It said, "with one's fingers and thumbs!"
Recollecting with tears how, in earlier years,
  It had taken no pains with its sums.

"The thing can be done," said the Butcher, "I think.
  The thing must be done, I am sure.
The thing shall be done! Bring me paper and ink,
  The best there is time to procure."

The Beaver brought paper, portfolio, pens,
  And ink in unfailing supplies:
While strange creepy creatures came out of their dens,
  And watched them with wondering eyes.

So engrossed was the Butcher, he heeded them not,
  As he wrote with a pen in each hand,
And explained all the while in a popular style
  Which the Beaver could well understand.

"Taking Three as the subject to reason about —
  A convenient number to state —
We add Seven, and Ten, and then multiply out
  By One Thousand diminished by Eight.

9

The Butcher instructs the Beaver

The result we proceed to divide, as you see,
  By Nine Hundred and Ninety and Two:
Then subtract Seventeen, and the answer must be
  Exactly and perfectly true."

The arithmetic described in this verse is straightforward. In trying to explain to the Beaver why 2 + 1 = 3, the Butcher starts with 3, adds 7 and 10, and multiplies by 1000 − 8 (which is 992). He then divides by 992 and subtracts 17, taking him back to where he started — namely, 3:

$$\frac{(3 + 7 + 10) \times (1000 - 8)}{992} - 17 = 3$$

In fact, any number other than 3 would have done equally well — the Butcher must always end with the number he started with.

Scene 6: Map-making

Earlier, in Fit the Second of *The Hunting of the Snark*, the Bellman provides a map for his crew of Snark hunters to use:

The Bellman himself they all praised to the skies —
  Such a carriage, such ease and such grace!
Such solemnity, too! One could see he was wise,
  The moment one looked in his face.

He had bought a large map representing the sea,
  Without the least vestige of land:
And the crew were much pleased when they found it to be
  A map they could all understand.

"What's the good of Mercator's North Poles and Equators,
  Tropics, Zones and Meridian Lines?"
So the Bellman would cry: and the crew would reply,
  "They are merely conventional signs!"

"Other maps are such shapes, with their islands and capes!
  But we've got our brave Captain to thank"
(So the crew would protest) "that he's bought *us* the best —
  A perfect and absolute blank!"

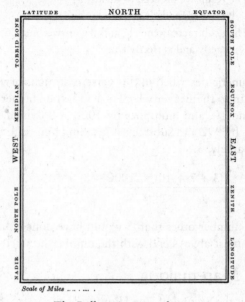

The Bellman's ocean chart

A different type of map is described in *Sylvie and Bruno Concluded* (1893), Lewis Carroll's last novel for children and the sequel to *Sylvie and Bruno* (1889). In this scene, the book's narrator (Myself) and the fairy children Sylvie and Bruno are listening to Mein Herr, a grand old German gentleman with a long beard, who explains to us how maps are constructed in his own country:

Myself: What a useful thing a pocket-map is!

Mein Herr: That's another thing we've learned from *your* Nation, map-making. But we've carried it much further than *you*. What do you consider the *largest* map that would be really useful?

Myself: About: six inches to the mile.

Mein Herr: Only *six inches*! We very soon got to six *yards* to the mile. Then we tried a *hundred* yards to the mile. And then

12

came the grandest idea of all! We actually made a map of the country, on the scale of *a mile to the mile*!

Myself: Have you used it much?

Mein Herr: It has never been spread out, yet: the farmers objected: they said it would cover the whole country, and shut out the sunlight! So we now use the country itself, as its own map, and I assure you it does nearly as well.

Scene 7: Fortunatus's Purse

In *Sylvie and Bruno Concluded*, Carroll's ability to illustrate mathematical ideas in a painless and picturesque way is shown in the construction of *Fortunatus's Purse* from three handkerchiefs. This purse has no inside or outside, and so can be considered to contain the entire wealth of the world.

The passage includes a description of a 'Paper Ring', or *Möbius band*, named after the nineteenth-century German mathematician and astronomer August Ferdinand Möbius. This can be made from a rectangular strip of paper by twisting one end through 180 degrees and then gluing the two ends together, as pictured here. The resulting object has just one side and just one edge: this means that an insect could travel from any point on it to any other point without leaving the surface or going over the edge.

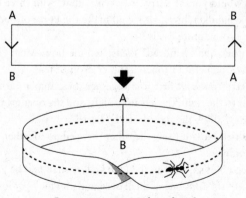

Constructing a Möbius band

An extension of this idea is to start with a rectangular strip and try to glue *both* pairs of opposite sides in opposing directions. This cannot be done in our three-dimensional world, however.

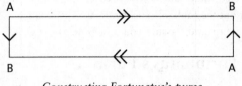

Constructing Fortunatus's purse

The resulting object — Fortunatus's Purse — has the form of a mathematical object called a *projective plane*. Since it cannot exist in three dimensions, the description that follows ceases just before the task becomes impossible.

We are in a shady nook where afternoon tea is being enjoyed. Lady Muriel is sewing, while her father (the Earl of Ainslie) and the narrator look on. Along comes the venerable Mein Herr.

Mein Herr: Hemming pocket-handkerchiefs? So *that* is what the English miladies occupy themselves with, is it?

Myself: It is the one accomplishment in which Man has never yet rivalled Woman!

Mein Herr: You have heard of Fortunatus's Purse, Miladi? Ah, so! Would you be surprised to hear that, with three of these leetle handkerchiefs, you shall make the Purse of Fortunatus quite soon, quite easily?

Lady Muriel: Shall I indeed? *Please* tell me how, Mein Herr! I'll make one before I touch another drop of tea!

Mein Herr: You shall first join together these upper corners, the right to the right, the left to the left; and the opening between them shall be the *mouth* of the Purse.

Lady Muriel: Now if I sew the other three edges together, the bag is complete?

Mein Herr: Not so, Miladi: the *lower* edges shall *first* be joined — ah, not so! Turn one of them over, and join the *right* lower corner of the one to the *left* lower corner of the other, and sew the lower edges together in what you would call *the wrong way*.

14

Lady Muriel: *I* see! And a very twisted, uncomfortable, uncanny-looking bag it makes! But the *moral* is a lovely one. Unlimited wealth can only be obtained by doing things *in the wrong way*! And how are we to join up these mysterious — no, I mean *this* mysterious opening? Yes, it *is* one opening — I thought it was *two*, at first.

Mein Herr: You have seen the puzzle of the Paper Ring? Where you take a slip of paper, and join its ends together, first twisting one, so as to join the *upper* corner of *one* end to the *lower* corner of the *other*?

The Earl: I saw one made, only yesterday. Muriel, my child, were you not making one, to amuse those children you had to tea?

Lady Muriel: Yes, I know that Puzzle. The Ring has only *one* surface, and only *one* edge. It's very mysterious!

Myself: The *bag* is just like that, isn't it? Is not the *outer* surface of one side of it continuous with the *inner* surface of the other side?

Mein Herr manipulates the handkerchiefs

15

Lady Muriel: So it is! Only it *isn't* a bag, just yet. How shall we fill up this opening, Mein Herr?

Mein Herr: Thus! The edge of the opening consists of *four* hand-kerchief-edges, and you can trace it continuously, round and round the opening: down the right edge of *one* handkerchief, up the left edge of the *other*, and then down the left edge of the *one*, and up the right edge of the *other*!

Lady Muriel: So you can! And that *proves* it to be only *one* opening!

Mein Herr: Now, this *third* handkerchief has *also* four edges, which you can trace continuously round and round: all you need do is to join its four edges to the four edges of the open-ing. The Purse is then complete, and its outer surface —

Lady Muriel: *I* see! Its *outer* surface will be continuous with its *inner* surface! But it will take time. I'll sew it up after tea. But why do you call it Fortunatus's Purse, Mein Herr?

Mein Herr: Don't you see, my child — I should say Miladi? Whatever is *inside* that Purse, is *outside* it; and whatever is *outside* it, is *inside* it. So you have all the wealth of the world in that leetle Purse!

Lady Muriel: I'll certainly sew the third handkerchief in — *some* time, but I wo'n't take up your time by trying it now.

Scene 8: A Question of Gravity

Still in our shady nook, Mein Herr reminisces about various inven-tions to be seen in his country, including a train that runs entirely by gravity.

Lady Muriel: *Please* tell us some more wonderful things!

Mein Herr: They run their railway-trains without any engines — nothing is needed but machinery to *stop* them with. Is *that* wonderful enough, Miladi?

Myself: But where does the *force* come from?

Mein Herr: They use the force of *gravity*. It is a force known also in *your* country, I believe?

The Earl: But that would need a railway going *down-hill*. You ca'n't have *all* your railways going down-hill?

Mein Herr: They *all* do.

16

The Earl: Not from *both* ends?

Mein Herr: From *both* ends.

The Earl: Then I give it up!

Lady Muriel: Can you explain the process?

Mein Herr: Easily. Each railway is in a long tunnel, perfectly straight: so of course the *middle* of it is nearer the centre of the globe than the two ends: so every train runs half-way *down*-hill, and that gives it force enough to run the *other* half *up*-hill.

Lady Muriel: Thank you. I understand that perfectly. But the velocity in the *middle* of the tunnel must be something *fearful*!

Gravity fascinated Lewis Carroll. *Alice's Adventures in Wonderland* commences with Alice tumbling down a deep rabbit-hole and wondering to herself how far she had fallen:

I wonder how many miles I've fallen by this time? I must be getting somewhere near the centre of the earth. Let me see: that would be four thousand miles down, I think . . . I wonder if I shall fall right *through* the earth! How funny it'll seem to come out among the people that walk with their heads downwards! The Antipathies, I think . . .

While descending, she takes a jar labelled ORANGE MARMALADE from a shelf and finds, to her great disappointment, that it is empty. She decides not to drop it for fear of killing anyone underneath, forgetting that it would remain suspended in front of her as she continued to fall.

This idea is developed further in *Sylvie and Bruno*, where Lady Muriel, her father the Earl and the narrator (Myself) are in conversation with a young doctor called Arthur. The narrator has just insisted on taking a cup of tea across the room to the Earl, and the conversation soon turns to the problem of drinking tea inside a falling house:

Lady Muriel: How convenient it would be if cups of tea had no weight at all! Then perhaps ladies would *sometimes* be permitted to carry them for short distances!

17

Arthur: One can easily imagine a situation where things would *necessarily* have no weight, relatively to each other, though each would have its usual weight, looked at by itself.

The Earl: Some desperate paradox! Tell us how it could be. We shall never guess it.

Arthur: Well, suppose this house, just as it is, placed a few billion miles above a planet, and with nothing else near enough to disturb it: of course, it falls *to* the planet?

The Earl: Of course — though it might take some centuries to do it.

Lady Muriel: And is five-o'clock-tea to be going on all the while?

Arthur: That, and other things. The inhabitants would live their lives, grow up and die, and still the house would be falling, falling, falling! But now as to the relative weight of things. Nothing can be *heavy*, you know, except by *trying* to fall, and being prevented from doing so. You all grant that?

All: Yes.

Arthur: Well, now, if I take this book, and hold it out at arm's length, of course I feel its *weight*. It is trying to fall, and I prevent it. And, if I let go, it falls to the floor. But, if we were all falling together, it couldn't be *trying* to fall any quicker, you know: for, if I let go, what more could it do than fall? And, as my hand would be falling too — at the same rate — it would never leave it, for that would be to get ahead of it in the race. And it could never overtake the falling floor!

Lady Muriel: I see it clearly. But it makes me dizzy to think of such things! How *can* you make us do it?

Myself: There is a more curious idea yet. Suppose a cord fastened to the house, from below, and pulled down by someone on the planet. Then of course the *house* goes faster than its natural rate of falling: but the furniture — with our noble selves — would go on falling at their old pace, and would therefore be left behind.

The Earl: Practically, we should rise to the ceiling. The inevitable result of which would be concussion of brain.

Arthur: To avoid that, let us have the furniture fixed to the floor, and ourselves tied down to the furniture. Then the five-o'-clock-tea could go on in peace.

18

Lady Muriel: With one little drawback! We should take the *cups* down with us: but what about the *tea*?

Arthur: I had forgotten the *tea*. *That*, no doubt, would rise to the ceiling — unless you chose to drink it on the way!

The Earl: Which, I think, is *quite* nonsense enough for one while!

Enough nonsense, indeed! After all these excursions into the world of his alter ego, Lewis Carroll, we now turn our attention to the early life of Charles Dodgson himself.

Charles Dodgson's England

Fit the First
The Children of the North

Lewis Carroll — or Charles Dodgson, as we must call him here — was born into a good English Church family. The third of eleven children, he was raised at Daresbury in Cheshire and Croft in Yorkshire, and went to boarding school in Richmond and Rugby before going up to Oxford University. In this chapter we outline the progress of these early years.

Daresbury

> An island farm, 'mid seas of corn,
> Swayed by the wandering breath of morn,
> The happy spot where I was born.

Charles Lutwidge Dodgson (pronounced 'dodson') was born on 27 January 1832 at the Old Parsonage in Newton-by-Daresbury, near the secluded village of Daresbury (pronounced 'darsbury') in Cheshire, where his father was perpetual curate. Here the young boy spent the first eleven years of his life deep in the countryside, where 'even the passing of a cart was a matter of great interest to the children'.

His father, the Reverend Charles Dodgson, was one of a long line of clergy stretching back several generations. A pious and deeply religious man for whom 'mathematics were his favourite pursuit', he enjoyed a brilliant early career at Westminster School, where he became Head Boy, and at Oxford University, where he received a double First Class degree in Classics and Mathematics at Christ Church in 1821. He was awarded a Studentship at Christ Church (more or less equivalent to a Fellowship in other colleges), which entitled him to live in College for the rest of his life, provided he remain unmarried and prepared for holy orders. He was ordained Deacon in 1823 and Priest the following year.

Two years later he determined to marry his first cousin, Fanny Lutwidge, 'one of the sweetest and gentlest women that ever lived,

The Revd Charles Dodgson
(photographed by Charles Dodgson, 1860)

Daresbury Parsonage
(photographed by Charles Dodgson, 1860)

whom to know was to love', and duly forfeited his Studentship. Christ Church presented him with a living at the parish church of All Saints, Daresbury, seven miles from Warrington and about twenty miles from Liverpool. The parsonage was one and a half miles from the village on a glebe farm, farmland that belonged to the parish and was let out for rent. In later years Charles photographed the parsonage, before it was destroyed by fire in 1884.

It was at Daresbury that the Revd Dodgson and his young wife started their large family of seven girls and four boys. After Charles's elder sisters (Fanny and Elizabeth) came Charles himself, followed by two more girls (Caroline and Mary), two more boys (Skeffington and Wilfred) and three more girls (Louise, Margaret and Henrietta). The youngest boy (Edwin) was born after the family had left Daresbury. All survived to adulthood.

Charles, as the eldest son, soon established himself as the children's natural leader, delighting in entertaining his ever-increasing family of brothers and sisters. In the isolated surroundings of Daresbury he derived great pleasure from the animal world around him, as he 'made pets of the most odd and unlikely animals, and numbered certain snails and toads among his intimate friends'.

The Reverend's meagre income of less than £200 per year, including what he earned from letting the glebe, was insufficient for his growing family's needs, and he supplemented it by taking private pupils. The parish of 146 parishioners had previously been somewhat inactive, but over sixteen years he carefully tended it, visiting the poor and needy, increasing the Sunday congregations, starting a Sunday school and instituting weekly lectures on a range of topics.

The Dodgson family received a strict Christian upbringing. Sunday was devoted solely to such activities as reading religious books, learning extended passages from the Bible, and attending morning and evening services at the church for their father's extempore sermons. Charles inherited a deep religious conviction and a sense of spirituality that would govern his future life.

The Dodgson parents educated their children at home. Charles, in particular, received from his father a thorough grounding in mathematics, Latin, Christian theology and English literature, subjects which would feature prominently throughout his life. Of his mathematical precocity, the story is told that

One day, when Charles was a very small boy, he came up to his father and showed him a book of logarithms, with the request, "Please explain." Mr. Dodgson told him that he was much too young to understand anything about such a difficult subject. The child listened to what his father said, and appeared to think it irrelevant, for he still insisted, "*But*, please, explain!"

Croft

Fair stands the ancient Rectory,
The Rectory of Croft,
The sun shines bright upon it,
The breezes whisper soft.
From all the house and garden
Its inhabitants come forth,
And muster in the road without,
And pace in twos and threes about,
The children of the North.

In 1836, in addition to his Daresbury duties, the Reverend Dodgson became Examining Chaplain to his old friend C.T. Longley, the Bishop of Ripon (later Archbishop of Canterbury). Seven years later the Crown appointment of the living at Croft-on-Tees, near the border between Yorkshire and Durham, became vacant, and the Bishop wrote to the Prime Minister, Sir Robert Peel, recommending Dodgson for the post.

Thus it was that in late 1843 the younger Charles Dodgson moved with his family to Croft, where his father became Rector of the parish church of St Peter's. The Rectory, just two minutes' walk from the church, was a large Georgian house with servants' rooms, adjacent farm buildings, and set in a large garden stocked with fruit trees and exotic flowering plants collected by the previous incumbent.

At Croft, Charles enjoyed an idyllic childhood with his brothers and sisters. There were delightful walks in the Yorkshire countryside and many games to play. He enjoyed writing and painting, and

Dodgson's sisters at Croft Rectory
(photographed by Charles Dodgson, 1862)

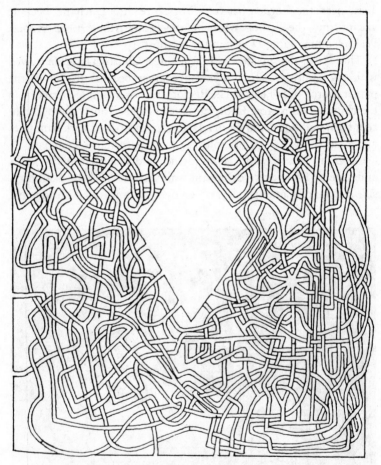

An intricate maze, designed by Charles Dodgson for the family magazine Mischmasch

derived much pleasure from organizing puppet shows with marionettes that he made himself, and entertaining the family with conjuring displays, arrayed in a brown wig and a long white robe. The railways were just arriving in Yorkshire, and Charles followed the fashion by constructing 'a rude train from a wheelbarrow, a barrel and a small truck, which used to convey passengers from one "station" in the Rectory garden to another'. One winter, he constructed a maze in the snow 'of such hopeless intricacy as almost to put its famous rival at Hampton Court in the shade'.

Shortly after arriving at Croft, Charles started a succession of family magazines, containing poems, sketches and other writings by himself and other members of the family. The first of these was *Useful and Instructive Poetry*, written by Charles when he was about thirteen for his brother Wilfred (aged seven) and sister Louisa (aged five). It included a short verse on astronomy, which became a lifelong interest:

> Were I to take an iron gun,
> And fire it off towards the sun;
> I grant t'would reach its mark at last,
> But not till many years had passed.
>
> But should that bullet change its' force,
> And to the planets take its' course,
> T'would *never* reach the *nearest* star,
> Because it is so *very* far.

Richmond

At Daresbury, the Reverend Dodgson's meagre income had required the parents to educate their children at home. With his move to Croft, his income increased to over £1000 per year. He could now afford to send Charles to a private school, to build his son's character and prepare him for the Church. In August 1844, Charles started at Richmond Grammar School, a school of 120 pupils just ten miles from Croft, where the fees were over £100 per year.

At Richmond School the curriculum consisted mainly of religious instruction and the classical languages and literature, with

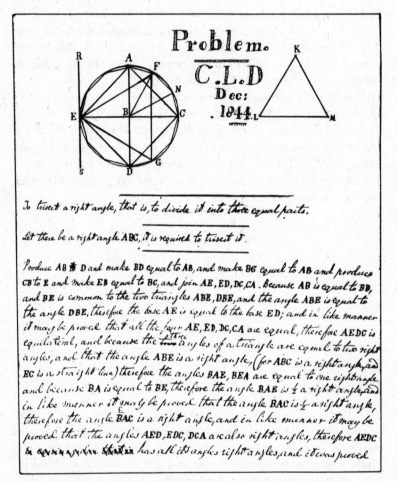

Problem.

C.L.D

Dec:
1844.

To trisect a right angle, that is, to divide it into three equal parts.

Let there be a right angle ABC, it is required to trisect it.

Produce AB to D and make BD equal to AB, and make BC equal to AB and produce CB to E and make EB equal to BC, and join AE, ED, DC, CA. Because AB is equal to BD, and BE is common to the two triangles ABE, DBE, and the angle ABE is equal to the angle DBE, therefore the base AE is equal to the base ED; and in like manner it may be proved that all the four AE, ED, DC, CA are equal, therefore AEDC is equilateral, and because the three angles of a triangle are equal to two right angles, and that the angle ABE is a right angle, (for ABC is a right angle, and EC is a straight line) therefore the angles BAE, BEA are equal to one right angle, and because BA is equal to BE, therefore the angle BAE is ½ a right angle, and in like manner it may be proved that the angle BAC is ½ a right angle, therefore the angle BAC is a right angle, and in like manner it may be proved that the angles AED, EDC, DCA are also right angles, therefore AEDC has all its angles right angles, and it was proved

A page of Charles's geometry, written when he was twelve

Richmond Grammar School

mathematics, French and accounting as optional extras. Charles received high marks and good reports, and the headmaster, James Tate, the 'kind old schoolmaster' with whom he boarded, enthused to Charles's father that

> he is capable of acquirements and knowledge far beyond his years, while his reason is so clear and so jealous of error, that he will not rest satisfied without a most exact solution of whatever seems to him obscure. He has passed an excellent examination just now in mathematics, exhibiting at times an illustration of that love of precise argument, which seems to him natural.

By this time Charles was composing Latin verse, and the page of geometry shown opposite demonstrates how far his mathematical interests and abilities had developed.

The mathematics textbook used at Richmond School was Francis Walkingame's *The Tutor's Assistant, being a Compendium of Arithmetic* in *Crosby's New Edition, with Considerable Additions*, of 1842. This classic eighteenth-century text went through scores of editions and contained arithmetical problems of a style that would be unacceptable today and of a complexity that would strike terror into the heart of many a present-day schoolchild; a selection of

Some problems from Francis Walkingame's arithmetic text

What is the cube root of 673373097125? *Ans. 8765.*

In an army consisting of 187 squadrons of horse, each 157 men, and 207 battallions, each 560 men — how many effective soldiers, supposing that in 7 hospitals there are 473 sick? *Ans. 144806.*

If from London to York be accounted 50 leagues; I demand how many miles, yards, feet, inches, and barleycorns? *Ans. 150 miles, 264000 yards, 792000 feet, 9504000 inches, 28512000 barleycorns.*

What sum did that gentleman receive in dowry with his wife, whose fortune was her wedding-suit: her petticoat having 2 rows of furbelows, each furbelow 87 quills, and each quill 21 guineas? *Ans. £3836: 14s.*

A gentleman going into a garden, meets with some ladies, and says to them, Good morning to you, 10 fair maids. Sir, you mistake, answered one of them, we are not 10; but if we were twice as many more as we are, we should be as many above 10 as we are now under — how many were they? *Ans. 5.*

If 100 eggs were placed in a right line, exactly a yard asunder from one another, and the first a yard from a basket, what length of ground does that man go who gathers up the hundred eggs singly, returning with every egg to the basket to put it in? *Ans. 5 miles, 1300 yards.*

If 504 Flemish ells, 2 qrs. cost 283 l. 17s. 6d.; at what rate must I give for 14 yards? *Ans. £10:10s.*

At 110¼ per cent. — what is the purchase of 2054 l. 16s. South-Sea stock? *Ans. £2265 : 8 : 4.*

A refiner having 12 lb. of silver bullion, of 6 oz. fine, would melt it with 8 lb. of 7 oz. fine, and 10 lb. of 8 oz. fine; I require the fineness of 1 lb. of that mixture? *Ans. 6 oz. 18 dwt. 16 gr.*

The spectators' club of fat people, though it consisted of 15 members, is said to have weighed no less than 3 tons — how much, at an equality, was that per man? *Ans. 4 cwt.*

Walkingame's problems appears opposite. Among its pages are methods for extracting cube roots, calculations on the purchasing of stock, and tables of measurements (many no longer in use), including those for ale and beer — topics of doubtful relevance for a twelve-year-old boy.

In March 1845 Charles inscribed his copy as follows:

> PRAEFATIO. Hic liber ad Carolum Ludrigum Dodsonum pertinet. O Lector! cave ne illum capias, nam latro Jovi est odius. Ecce!
> [*PREFACE. This book belongs to Charles Lutwidge Dodgson. O reader! Take care not to steal it, for a thief is odious to Jove. Behold!*]

This Latinization of his name predated by eleven years his later pen-name of Lewis Carroll.

When Charles left Richmond School, his headmaster wrote again to his father:

> Be assured that I shall always feel a particular interest in the gentle, intelligent, and well-conducted boy who is now leaving us.

Rugby

In February 1846, the fourteen-year-old Charles Dodgson was sent to Rugby School, where he delighted in his studies of mathematics and the Classics, but suffered from the rougher elements. Like other new boys he was subjected to bullying, and winter nights in the dormitories were intolerably cold, as the older boys kept themselves warm by stripping the bedclothes from the younger ones. His health also suffered. In 1848 he developed a severe case of whooping cough, and later contracted mumps, aggravating the deafness in his right ear that had developed some years earlier.

In later years Charles looked back on his Rugby days with distaste:

> During my stay I made I suppose some progress in learning of various kinds, but none of it was done *con amore*, and I spent an incalculable time in writing out impositions — this last I consider one of the chief faults of Rugby School. I made some friends there . . . but I cannot say that I look back upon my life at a Public School with any sensations of pleasure, or that any earthly considerations would induce me to go through my three years again.

Rugby School

And following a visit to another school, where the boys were assigned separate sleeping areas, he wrote:

> I can say that if I had been thus secure from annoyance at night, the hardships of the daily life would have been comparative trifles to bear.

The teaching and curriculum at Rugby were traditional. Each week the instruction, which started at 7 a.m., consisted of sixteen lengthy classes in the Classics, history and Divinity, but only two in French and two in mathematics, and none in science. In spite of this, Charles made good progress with his studies, being regarded as exceptionally gifted in mathematics and winning prizes for history, Divinity, mathematics, Latin composition and English. At the end of his first year he came first in mathematics in the lower fifth form, receiving Conyers Middleton's *Life of Cicero* from the Masters of Rugby School for his award. A year later he won the 2nd Mathematical Prize in the annual general examination and the '1st Prize of the division in Mathematics for the work of the half', and was presented with William Roscoe's *Life of Lorenzo di Medici*. Around Christmas 1848 he achieved first class in mathematics and other subjects, receiving Leopold von Ranke's *History of the Popes*. By this means he was gradually building up his personal library. Indeed, one of his prize purchases while at Rugby was Liddell and Scott's larger *Greek–English Lexicon*; he could not have foreseen the extent to which the Liddell family would affect his later life.

In a letter to his sister Elizabeth, written in October 1848, he commented on his school report and on some of the other Prizemen:

The report is certainly a delightful one: *I* cannot account for it; I hope there is no mistake. As to the difference between Walker and myself . . . it must be remembered that he is in the 6th and has hitherto been considered the best mathematician in the school. Indeed no one but me got anything *out* of the 6th . . . As to the tutor marks, we did not go the 1st week and the Prize examinations have prevented the 4th. The Lower Mathematical Prizeman, Fisher, unfortunately broke his arm yesterday by falling down.

A year later, in another letter to Elizabeth, he observed that:

Yesterday evening I was walking out with a friend of mine who attends as mathematical pupil Mr. Smythies the second mathematical master; we went up to Mr. Smythies' house, as he wanted to speak to him, and he asked us to stop and have a glass of wine and some figs. He seems as devoted to his duty as Mr. Mayor, and asked me with a smile of delight, "Well, Dodgson, I suppose you're getting well on with your mathematics?" He is very clever at them, though not equal to Mr. Mayor, as indeed few men are, Papa excepted.

His teachers were enthusiastic about his progress. In 1848 his mathematics master, Robert Mayor, confided to the Reverend Dodgson that:

I have not had a more promising boy at his age since I came to Rugby.

And just before Charles left Rugby School, the headmaster, Dr Archibald Tait (later Archbishop of Canterbury), wrote:

I must not allow your son to leave school without expressing to you the very high opinion I entertain of him. I fully coincide in Mr. Cotton's estimate both of his abilities and upright conduct. His mathematical knowledge is great for his age, and I doubt not he will do himself credit in classics. As I believe I mentioned to you before, his examination for the Divinity prize was one of the most creditable exhibitions I have ever seen.

Marking Time

After the hardships of Rugby School, Charles returned to Croft Rectory for a few months to prepare himself for the next stage of his career. He continued to produce family magazines for his brothers and sisters, and the 1849 issue of *The Rectory Umbrella* (named after the 'umbrella tree', a huge yew tree in the garden) contained two mathematical 'Difficulties'. They both involved the measurement of time, a topic that was to fascinate him throughout his life.

Difficulty No. 1

The first of the time problems concerned the rotation of the earth, and was later entitled 'Where does the day begin?' and 'A hemispherical problem':

> Half of the world, or nearly so, is always in the light of the sun: as the world turns round, this hemisphere of light shifts round too, and passes over each part of it in succession.
>
> Supposing on Tuesday, it is morning at London; in another hour it would be Tuesday morning at the west of England; if the whole world were land we might go on tracing Tuesday Morning, Tuesday Morning all the way round, till in 24 hours we get to London again. But we *know* that at London 24 hours after Tuesday morning it is Wednesday morning. Where then, in its passage round the earth, does the day change its name? where does it lose its identity? . . . some line would have to be fixed, where the change should take place, so that the inhabitant of one house would wake and say 'heigh-ho! Tuesday morning!' and the inhabitant of the next, (over the line,) a few miles to the west would wake a few minutes afterwards and say 'heigh-ho! Wednesday morning!' What hopeless confusion the people who happened to live *on* the line would be in, it is not for me to say.

This problem resurfaced in 1857 in the columns of *The Illustrated London News*. Dodgson joined the discussions, attempting to explain to the readers the source of the difficulty, and describing it as 'a question which occurred to myself years ago, and for which I have never been able to meet with a satisfactory solution'. In November 1860 he presented the problem again at a meeting of the Ashmolean

Society of Oxford, and years later it was noted that 'The difficulty of answering this apparently simple question has cast a gloom over many a pleasant party.' The matter was eventually resolved in 1884 with the establishment of the International Date Line.

The rotation of the earth also featured in his later writings. In *Alice's Adventures in Wonderland*, the following exchange takes place between Alice and the Duchess:

> **Duchess:** If everybody minded their own business, the world would go round a deal faster than it does.
> **Alice:** Which would *not* be an advantage. Just think what work it would make with the day and night! You see the earth takes twenty-four hours to turn round on its axis —
> **Duchess:** Talking of axes, chop off her head!
> **Alice:** Twenty-hour hours, I *think*; or is it twelve? I —
> **Duchess:** Oh, don't bother *me*! I never could abide figures!

Alice confronts the Duchess about time

Alice later encountered time at the famous mad tea party with the Hatter, the March Hare and the Dormouse. Following some confusion over the Hatter's watch, which told the day of the month rather than the time of day, Alice challenged him:

> **Alice:** I think you might do something better with the time than wasting it in asking riddles that have no answers.
>
> **Hatter:** If you knew Time as I do, you wouldn't talk about wasting *it*. It's *him*.
>
> **Alice:** I don't know what you mean.
>
> **Hatter:** Of course you don't! I dare say you never even spoke to Time!
>
> **Alice:** Perhaps not, but I know I have to beat time when I learn music.
>
> **Hatter:** Ah! That accounts for it. He won't stand beating. Now, if you only kept on good terms with him, he'd do almost anything you liked with the clock. For instance, suppose it were nine o'clock in the morning, just time to begin lessons: you'd only have to whisper a hint to Time, and round goes the clock in a twinkling! Half-past one, time for dinner!

Difficulty No. 2

Charles's fascination with time had also surfaced in a letter he sent to Elizabeth from Rugby School:

> Will you answer my question about the clocks, when you next write?

This 'question about the clocks' was the second of his two Difficulties, later called 'The Two Clocks'. It is simpler than the previous one, and we present it here in full. It is written in the whimsical style that would become so familiar to his readers in later years.

> Which is the best, a clock that is right only once a year, or a clock that is right twice every day? "The latter," you reply, "unquestionably." Very good, reader, now attend.
>
> I have two clocks: one doesn't go *at all*, and the other loses a minute a day: which would you prefer? "The losing one," you answer, "without a doubt." Now observe: the one which loses a minute a day has to lose twelve hours, or seven hundred and twenty

minutes before it is right again, consequently it is only right once in two years, whereas the other is evidently right as often as the time it points to comes round, which happens twice a day. So you've contradicted yourself *once*. "Ah, but," you say, "what's the use of its being right twice a day, if I can't tell when the time comes?" Why, suppose the clock points to eight o'clock, don't you see that the clock is right *at* eight o'clock? Consequently when eight o'clock comes your clock is right. "Yes, I see *that*," you reply. Very good, then you've contradicted yourself *twice*: now get out of the difficulty as best you can, and don't contradict yourself again if you can help it.

A footnote was added:

You *might* go on to ask, "How am I to know when eight o'clock *does* come? My clock will not tell me." Be patient, reader: you know that when eight o'clock comes your clock is right: very good; then your rule is this: keep your eye fixed on your clock, and *the very moment it is right* it will be eight o'clock. "But —" you say. There, that'll do, reader; the more you argue, the farther you get from the point, so it will be as well to stop.

But the time had now come for Charles Dodgson to put these schoolboy difficulties behind him, and head for Oxford.

Christ Church, Oxford, around 1850

Fit the Second
Uppe toe mine Eyes yn Worke

With his impressive Rugby School record, Charles Dodgson was well prepared to study at Oxford University. On 23 May 1850 he travelled to Oxford for his matriculation examinations in Latin, Greek and mathematics, and the ensuing ceremony where he pledged allegiance to the Thirty-Nine Articles of the Church of England and was officially enrolled as a member of the University.

At that time, Oxford was a small country town with unpaved streets and horse-drawn carriages. Then, as now, the University consisted of a number of colleges, and Charles became a member of Christ Church, the college where his father had achieved great success thirty years earlier.

Founded in the sixteenth century by Cardinal Wolsey, King Henry VIII's Chancellor, Christ Church includes the Cathedral of the Diocese of Oxford. It also features Christopher Wren's gate-tower, known as Tom Tower, and the Great Quadrangle, which, according to Dodgson, 'very vulgar people call "Tom Quad"'. During the English Civil War of the 1640s, King Charles I lived in Christ Church and held Parliamentary meetings in the magnificent Great Hall, where Dodgson later claimed to have 'dined several times (about 8000 times, perhaps)'. Every evening, Great Tom, the bell in the gate-tower, still rings 101 times to celebrate the 101 Students who became part of the College establishment in the seventeenth century.

An Oxford Undergraduate

Here's to the Freshman of bashful eighteen!
Here's to the Senior of twenty!
Here's to the youth whose moustache ca'n't be seen!
And here's to the man who has plenty!
Let the men Pass!
Out of the mass
I'll warrant we'll find you some fit for a Class!

39

When Charles Dodgson entered Christ Church for the first time, at the age of eighteen, he could hardly have expected that it would be his home for the rest of his life. He did not immediately take up residence, however, but returned to Croft Rectory to prepare himself for the start of his course. A family friend, Dr Jelf, one of the Canons of Christ Church, wrote to Charles's father:

> I am sure that I express the common feeling of all who remember you at Christ Church when I say that we shall rejoice to see a son of yours worthy to tread in your footsteps.

Charles's return to Oxford on 24 January 1851 was short-lived. Two days later his beloved mother, Fanny, died suddenly and unexpectedly of 'inflammation of the brain' at the age of only forty-seven, and he had to return home. This event was devastating for Charles, and also for his father, who needed to arrange for the care of his large family. After a short period, Fanny's sister, Lucy Lutwidge, arrived at Croft Rectory to care for the Dodgson children; Aunt Lucy remained with the family for the rest of her life.

Back in College, Charles settled into the routine of undergraduate life. He was called every morning at around 6.15, had breakfast in his rooms, attended College chapel at eight o'clock, and studied throughout the morning, wearing his gown and mortar board for lectures and tutorials and around the town. In the afternoons, after a light lunch, he relaxed — going for long walks in his greatcoat, grey gloves and silk top hat, boating with friends on the river in his white flannel trousers and straw hat, or watching a game of cricket. After dinner in the Great Hall at 5 p.m., Charles often spent his evenings reading in his rooms or composing letters, standing at his writing desk. Although several Christ Church undergraduates were noblemen from wealthy families who spent much of their time in riotous living, such as hunting, gaming and drinking, Charles, like his father thirty years earlier, was there for the purpose of serious study and the passing of examinations.

The University year was divided into four terms. Dodgson's first term was Hilary (or Lent) Term, from January until the end of March. This was followed by two short terms, Easter Term from

The Great Hall of Christ Church

late April until early June, and Trinity (or Act) Term from mid-June to early July. The Long Vacation extended for three months, to be followed by Michaelmas Term, which lasted from October until December.

Undergraduates could choose to be *passmen*, working for a Pass degree, which took about three years, or *classmen*, working for an Honours degree, which took an extra year. For Honours they had to pass in two subjects: *Literae Humaniores* (Classics), which was obligatory, and one of mathematics, natural science, or law and modern history. Christ Church required all its undergraduates to read for Honours, and Charles elected to work towards the four-year degree in Classics and mathematics.

At Oxford the teaching was carried out by the professors, who delivered university lectures; by college lecturers, who lectured to small groups of undergraduates; and by tutors, who gave private tuition. The Mathematical Lecturer at Christ Church was the newly appointed Robert Faussett, who taught Dodgson throughout his undergraduate career and became a close friend.

Robert Faussett
(photographed by Charles Dodgson, 1857)

A Trio of Examinations

A long table, covered with books . . . Two gloomy-browed examiners,
and twelve pale-faced youths.

Dodgson's university course required him to sit three main examinations involving written papers and viva voce examinations —
Responsions, Moderations and Finals.

Responsions
Responsions (known as 'Little-go') was the first hurdle that all
undergraduates had to overcome. It took place twice a year and

consisted of papers on the Latin and Greek authors, a translation from English into Latin, a paper of grammatical questions, basic arithmetic (up to the extraction of square roots) and a paper on algebra or geometry. Most attempted it after a year or more, but Dodgson was better prepared than most and took it in Trinity Term 1851.

On 10 June, shortly before starting his examinations, he wrote in mock Olde Englishe to his sister Louisa:

> My beloved and thrice-respected sister,
>   ... Ytte will pain thyne hearte, I wotte, toe heare thatte ye people offe Oxford hig-towne cannotte skylle to nurse babys; and trulie their mannere thereoffe is cruelle: herewithe I enclose a sketch of what I have wytnessed myne selfe, and to mie mynde the underneathe babie yn the nurse herre armes ys yn ane sorrie plight.
>   Onne Moone his daye nexte we goe yn forre Responsions, and I amme uppe toe mine eyes yn worke.
>   Thine truly, Charles.

His 'worke' paid dividends, and he passed. Shortly afterwards, in early July, he travelled to London with his Aunt Charlotte to visit the Great Exhibition of 1851 at the Crystal Palace. He was greatly excited by the variety of displays from around the world:

> I think the first impression produced on you when you get inside is of bewilderment. It looks like a sort of fairyland. As far as you can look in any direction, you see nothing but pillars hung about with shawls, carpets, &c with long avenues of statues, fountains, canopies &c &c &c.

Moderations

Dodgson spent the long summer vacation back at Croft Rectory with his brothers and sisters. In October he returned to Oxford to begin preparations for the second part of his examinations a year later, attending lectures on pure geometry by the Savilian Professor of Geometry (the Revd Baden Powell) and studying with the College Lecturer. For his continued progress, the College awarded Dodgson a scholarship of £20 per year.

The First Public Examination under Moderators took place in November 1852. It consisted of papers on the four Gospels in Greek, and on one Greek and one Latin author (of which one had to be a poet and the other an orator), and either a logic paper or one on geometry and algebra. In addition, candidates for Honorary Distinction in *Disciplinis Mathematicis* were required to take a paper in pure mathematics.

In order to prepare for these examinations, Dodgson had asked Robert Faussett and Osborne Gordon (the Classical moderator) what he should study over the Long Vacation prior to the examination:

> I believe 25 hours' *hard* work a day *may* get through all I have to do, but I am not certain.

But his mathematical studies had to wait. His summer vacation commenced with a visit to his maternal uncle, Skeffington Lutwidge, a barrister and Commissioner in Lunacy living in London, to admire his latest scientific gadgets.

> He has as usual got a great number of new oddities, including a lathe, telescope stand, crest stamp . . . a beautiful little pocket instrument for measuring distances on a map, refrigerator &c. &c. We had an observation of the moon and Jupiter last night, & afterwards live animalcula in his large microscope.

Fortunately, Dodgson's twenty-five hours of hard work per day paid off: he achieved a First Class in Mathematics and a Second Class in Classics. Overjoyed, he wrote to his sister Elizabeth:

> You shall have the announcement of the last piece of good fortune this wonderful term has had in store for me, that is, *a 1ˢᵗ. class in*

Mathematics. Whether I shall add to this any honours at Collections [a College examination] I cannot at present say, but I should think it very unlikely, as I have only today to get up the work in The Acts of the Apostles, 2 Greek Plays, and the Satires of Horace & I feel myself almost totally unable to read at all: I am beginning to suffer from the reaction of reading for Moderations ... I am getting quite tired of being congratulated on various subjects: there seems to be no end of it. If I had shot the Dean, I could hardly have had more said about it.

A College Studentship for his son had long been an ambition of Charles Dodgson, Senior. While his son was still at Rugby School, the Reverend had written to his friend Dr Pusey, one of the Canons of Christ Church, expressing the hope that if the young Charles reached the appropriate standard he would be duly rewarded. Pusey, never one for favouritism, replied:

I have now, for nearly twenty years, not given a Studentship to any friend of my own, unless there was no very eligible person in the College. I have passed by or declined the sons of those to whom I was personally indebted for kindness. I can only say that I shall have *very great* pleasure, if circumstances permit me to nominate your son.

An Oxford viva voce examination

Canon Pusey in Tom Quad

By long tradition, the six Canons met every year on Christmas Eve to restore the total number of Studentships to 101. As a result of Charles's success in his Moderations exams, Pusey was able to write again to the Reverend Dodgson:

> I have great pleasure in telling you that I have been enabled to recommend your son for a Studentship this Christmas. It must be so much more satisfactory to you that he should be nominated thus, in consequence of the recommendations of the College. One of the Censors brought me to-day five names; but in their minds it was plain that they thought your son on the whole the most eligible for the College. It has been very satisfactory to hear of your son's steady and good conduct.

Charles thus became a Student of Christ Church, as his father had been a generation earlier. This entitled him to reside in College, provided he remained celibate and prepared for holy orders — conditions that he took very seriously. Many years later, in a letter to his cousin and godson, he recalled how:

the Studentships at Christ Church were in the gift of the Dean and Chapter — each Canon having a turn: and Dr. Pusey, having a turn, sent for me, and told me he would like to nominate me, but had made a rule to nominate *only* those who were going to take Holy Orders. I told him that was my intention, and he nominated me.

His father was delighted, writing to Charles:

The feelings of thankfulness and delight with which I have read your letter just received, I must leave to *your conception*; for they are, I assure you, beyond *my expression*; and your affectionate heart will derive no small addition of joy from thinking of the joy which you have occasioned to me, and to all the circle of your home. I say "*you* have occasioned," because, grateful as I am to my old friend Dr. Pusey for what he has done, I cannot desire stronger evidence than his own words of the fact that you have *won*, and well won, this honour for *yourself*, and that it is bestowed as a matter of *justice* to *you*, and not of *kindness* to *me*.

It was indeed a busy time for the Reverend Dodgson. In late 1852 he had been appointed a Canon of Ripon Cathedral, requiring him to spend the first three months of each year in Ripon before returning home to Croft Rectory for the other nine months. Soon afterwards he would also be appointed Archdeacon of Richmond.

Finals

The Public Examination of Finals was the culmination of Charles's undergraduate career, and consisted of two parts. The first of these was Greats, in the Easter Term of 1854. It tested the classical languages and literature, together with ancient history and philosophy, and was compulsory for everyone. In spite of working thirteen hours a day for the three weeks before the examination, and spending the whole night before the viva voce examination over his books, Charles emerged with only a Third Class.

The second part of his Finals degree was in his chosen area of mathematics, for which the minimum requirement was 'The first six books of Euclid, or the first part of Algebra'. Candidates for Honours were also required to study 'Mixed as well as Pure Mathematics', which covered a range of topics from the differential and integral calculus to astronomy and optics.

In order to prepare for these examinations, Dodgson spent much of his summer vacation on a two-month mathematical reading party at Whitby in Yorkshire. This was led by Bartholomew Price, the recently appointed Sedleian Professor of Natural Philosophy and the author of distinguished treatises on the calculus. His first name was frequently abbreviated to 'Bat', and appears in the Hatter's parody of 'Twinkle, Twinkle, Little Star':

> Twinkle, twinkle, little bat!
> How I wonder what you're at!
> Up above the world you fly.
> Like a tea-tray in the sky.

Dodgson developed a great admiration for Professor Price, who would later become a close friend and adviser. From Whitby he confided to his sister Mary that

I am doing Integral Calculus with him now, and getting on very swimmingly.

It was not all work at the reading party: there was time also for recreational activities. Dodgson wrote a story and a poem for the local newspaper, and many years later one of the other participants, Thomas Fowler (later President of Corpus Christi College), recalled that

Dodgson and I were both pupils of Professor Bartholomew Price (now Master of Pembroke) in a mathematical Reading Party at Whitby in the summer of 1854. It was there that "Alice" was incubated. Dodgson used to sit on a rock on the beach, telling stories to a circle of eager young listeners of both sexes.

Although the reference to Alice must be incorrect, since Charles was not yet aware of her existence, he was already entertaining young children with his stories.

Almost thirty years later, Charles enjoyed a reunion with two of his Whitby friends:

Dined with Fowler (now President of C.C.C.) in hall, to meet Ranken. Both men are now mostly bald, with quite grey hair: yet how short a time it seems since we were undergraduates together at Whitby!

Bartholomew 'Bat' Price

It was also many years later, in a letter to a young friend, that he recalled his mathematical studies at Oxford and gave some useful hints for studying:

When I was reading Mathematics for University honours, I would sometimes, after working a week or two at some new book, and mastering ten or twenty pages, get into a hopeless muddle, and find it just as bad the next morning. My rule was *to begin the book*

SECOND PUBLIC EXAMINATION.

I.

Geometry and Algebra.

1. Compare the advantages of a decimal and of a duo-decimal system of notation in reference to (1) commerce, (2) pure arithmetic; and shew by duodecimals that the area of a room whose length is 29 feet $7\frac{1}{4}$ inches, and breadth is 33 feet $9\frac{1}{4}$ inches, is 704 feet $30\frac{2}{5}$ inches.

2. Planes which are perpendicular to parallel straight lines are parallel to one another: and all planes which cut orthogonally a given circle meet in one and the same straight line.

3. Solve the following equations:

(1) $\dfrac{x + \sqrt{a^2 - x^2}}{x - \sqrt{a^2 - x^2}} = b.$

(2) $\left.\begin{array}{l} x^3 - y^3 = 98 \\ x - y = 2 \end{array}\right\}$

(3) $\left.\begin{array}{l} \dfrac{x}{a} + \dfrac{y}{b} = 1 \\[2mm] \dfrac{z}{c} + \dfrac{x}{a} = 1 \\[2mm] yz = bc \end{array}\right\}$

4. The difference of the squares of any two odd numbers is divisible by 8.

5. Shew that in a binomial, (whose index is a positive whole number,) the coefficient of any term of the expansion reckoned from the end, is the same as the coefficient of the corresponding term reckoned from the beginning.

6. In a given equilateral triangle a circle is inscribed, and then in the triangle formed by a tangent to that circle parallel to any side and the parts of the original triangle cut off by it, another circle is inscribed, and so on *ad infinitum*. Find the sum of the radii of these circles.

[Turn over.

A Finals paper in mathematics from the 1850s

again. And perhaps in another fortnight I had come to the old difficulty with impetus enough to get over it. Or perhaps not. I have several books that I have begun over and over again.

My second hint shall be — Never leave an unsolved difficulty
behind. I mean, don't go any further in that book till the difficulty is
conquered. In this point, Mathematics differs entirely from most
other subjects. Suppose you are reading an Italian book, and come
to a hopelessly obscure sentence — don't waste too much time on it,
skip it, and go on; you will do very well without it. But if you skip
a *mathematical* difficulty, it is sure to crop up again: you will find
some other proof depending on it, and you will only get deeper and
deeper into the mud.

My third hint is, only go on working so long as the brain is *quite*
clear. The moment you feel the ideas getting confused leave off and
rest, or your penalty will be that you will never learn Mathematics
at all!

Charles's Finals examinations in mathematics took place in
October and November 1854, and ranged over many areas of
pure and applied mathematics. The examiners were Bartholomew
Price, William Donkin (the Savilian Professor of Astronomy) and
Henry Pritchard of Corpus Christi College, and the examination
resulted in a Class list of five First Class degrees, seven Seconds, no
Thirds, six Fourths, and thirty-five Pass degrees (listed as Fifth Class).
Charles Dodgson was extremely successful, coming top of the
entire list. In a letter to his sister Mary, he enthused:

My dear Sister,
Enclosed you will find a list which I expect you to rejoice over considerably; it will take me more than a day to believe it, I expect — I
feel at present very much like a child with a new toy, but I daresay I
shall be tired of it soon, and wish to be Pope of Rome next. Those
in the list who were in the Whitby party are, Fowler, Ranken,
Almond and Wingfield. I have just given my Scout a bottle of wine
to drink to my First. We shall be made Bachelors on Monday . . . I
have just been to Mr. Price to see how I did in the papers, and the
result will I hope be gratifying to you. The following were the sum

An Oxford graduation ceremony

total of the marks for each in the First Class, as nearly as I can remember:–

Dodgson 279
Bosanquet 261
Cookson 254
Fowler 225
Ranken 213

He also said he never remembered so good a set of men in. All this is very satisfactory.

Your very affectionate brother,
Charles L. Dodgson

Charles Dodgson received his Bachelor of Arts degree at the graduation ceremony on 18 December 1854, bringing his undergraduate days to a triumphant close.

Fit the Third
Successes and Failures

With his Finals examinations safely behind him, Dodgson returned to Oxford in early 1855 to resume life as a Student at Christ Church. Robert Faussett was about to give up his lectureship and leave Oxford to take up an army commission fighting in the Crimean War. In his celebratory letter to his sister Mary, the new graduate mused about his prospects for the coming year:

> I must also add (this is a very boastful letter) that I ought to get the senior scholarship next term. Bosanquet will not try, as he is leaving Oxford, and the only man, beside the present first, to try, is one who got a 2nd last time. One thing more I will add, to crown all, and that is, I find I am the next First Class Mathematical Student to Faussett (with the exception of Kitchin who has given up Mathematics), so that I stand next (as Bosanquet is going to leave) for the Lectureship. And now I think that is enough news for one post.

Preparing for the scholarship and the lectureship was to consume much of his time and attention, but there were other preoccupations, as we shall see.

The Senior Scholarship

Every year, Oxford University awarded Junior and Senior Mathematical Scholarships by examination, the latter being usually taken after Finals, and Dodgson resolved to study for the Senior Scholarship examinations in March. However, after the exertions of his Finals examinations, his Christmas vacation with the family in Croft and at Ripon saw him dabbling at his mathematical studies in a somewhat half-hearted and disorganized fashion. Just before returning to Oxford in mid-January, he recorded in his diary:

> Did a little Mathematics in the morning: however, the work done this Vacation must count as "nil."

See unfading in honours, immortal in years,
The great Mother of Churchmen and Tories appears.
New Oxford Sausage

In order to improve his chances of winning the scholarship, he arranged for regular coaching from Professor Price, whose company he had so enjoyed during the previous summer's reading party in Whitby. However, the work proved to be less straightforward than he had expected:

> I talked over the "Calculus of Variations" with Price today, but without any effect. I see no prospect of understanding the subject at all.

Additionally, he found he was having to concentrate more on college teaching. He was spending about fifteen hours each week teaching individual pupils, which left little time to prepare for the scholarship.

The inevitable happened. In March the scholarship examinations came and went, as he confessed in his diary:

> 22 *March*: Examination began for the Mathematical Scholarships: I only succeeded in doing five questions in the morning, and four in the afternoon.
>
> 23 *March*: I did only two questions in the morning paper, and accordingly gave up, and did not go in for the afternoon one, but went a long walk with Mayo instead ... The scholarship papers seem such as I ought to be able to do with a year's more reading, and I am in very good hopes of getting it next time.

The Senior Scholarship was duly awarded to Samuel Bosanquet, whom Dodgson had beaten into second place in Finals three months previously and who had unexpectedly remained in Oxford:

> It is some consolation to me to think that at least Ch. Ch. has got it. He tells me he did not do more than six questions in any one of the papers. Almond, and Norton, and myself, all gave up on the second day, so that Bosanquet and Cookson had the last paper all to themselves.

But he was clearly ashamed of his lack of success:

> It is tantalising to think how easily I might have got it, if only I had worked properly during this term, which I fear I must consider as wasted. However, I have now got a year before me ... I mean to have read by next time, Integral Calculus, Optics (and theory of light),

Astronomy, and higher Dynamics. I record this resolution to shame myself with, in case March/56 finds me still unprepared, knowing how many similar failures there have been in my life already.

College Teaching

At the beginning of term he had been approached by the Senior Censor to instruct a freshman who was preparing for Responsions:

Had my first interview with Burton, my first pupil: he seems to take in Algebra very readily. I doubt if it will be worth his while to coach two terms merely for his Little-Go — another lesson or two will decide.

The next day he outlined his approach to teaching in a whimsical letter to his sister Henrietta and brother Edwin:

My one pupil has begun his work with me, & I will give you a description how the lecture is conducted — It is the most important point, you know, that the tutor sh^{ld} be *dignified* & at a distance from the pupil, and that the pupil sh^{ld} be as much as possible *degraded* — Otherwise you know, they are not humble enough — So I sit at the further end of the room; outside the door, (*which is shut*) sits the scout; outside the outer door (*also shut*) sits the sub-scout; halfway down stairs sits the sub-sub-scout; and down in the yard sits the *pupil*.

    The questions are shouted from one to the other, and the answers come back in the same way, — it is rather confusing till you are well used to it — The lecture goes on, something like this:

Tutor "What is twice three?"
Scout "What's a rice tree?"
Sub-Scout "When is ice free?"
Sub-sub-Scout "What's a nice fee?"
Pupil (timidly) "Half a guinea!"
Sub-sub-scout "Can't forge any!"
Sub-scout "Hoe for Jinny!"
Scout "Don't be a ninny!"
Tutor (looks offended, but tries another question.) "Divide a hundred by twelve!"
Scout "Provide wonderful bells!"

Sub-scout "Go ride under it yourself!"
Sub-sub-scout "Deride the dunder-headed elf!"
Pupil (surprised) "Who do you mean?"
Sub-sub-scout "Doings between!"
Sub-scout "Blue is the screen!"
Scout "Soup-tureen!"

& so the lecture proceeds.
  Such is life — from y^r. most affc^{te} brother
  Charles L. Dodgson

Dodgson's approach to examinations could be equally fanciful, as the following 'proposition' illustrates:

> To find the value of a given Examiner.
> *Example*.– A takes in ten books in the Final Examination, and gets a 3rd Class. B takes in the Examiners, and gets a 2nd. Find the value of the Examiners in terms of books. Find also their value in terms in which no Examination is held.

Gradually he began to take on other pupils:

> Got a note from Leighton, a gentleman commoner, who wishes to be taught some arithmetic for his Little-Go (which comes in about a fortnight) — as well as the second book of Euclid. We settled that he is to come an hour an evening, and began at once: so far as we went, he is well up, and needs no teaching: he makes my third pupil.

Two weeks later, Baldwin Leighton passed his Responsions, and on receiving the £5 payment, Dodgson recorded that this was the first money he had ever earned.

By the Easter Term, the number of his private pupils had increased dramatically — to fourteen. Although not an official arrangement, Dodgson considered that the experience so gained would increase his chances of getting the Mathematical Lectureship, as well as bringing in about £50. He organized his pupils into tutorial groups, remarking ruefully that the fifteen hours of teaching each week would be 'a remedy against idleness, such as I could never have devised for myself'. The following list from Dodgson's diary shows the subjects he taught and the pupils taking each one:

Charles Dodgson as a young don

Differential Calculus: Blackmore.
Conics beginning: Burges, Wickham Senior.
Conics half-way: Corbet, Parham, Rattle.
Trigonometry beginning: Bellett, Lavie, Wickham Junior.
Trigonometry half-way: Paxton, Vesey.
Euclid and Algebra: Harington, Willatts.

Financially, Dodgson was now managing to stand on his own feet. He had been appointed College Sub-Librarian in February, which brought in £35 per year, and in May he recorded that

> The Dean and Canons have been pleased to give me one of the "Bostock" Scholarships — said to be worth £20 a year — this very nearly raises my income this year to independence. Courage!

A New Appointment

> I am the Dean, and this is Mrs Liddell:
> She plays the first, and I the second fiddle.
> She's the Broad and I'm the High:
> We're the University.

In June 1855, the Dean of Christ Church died and his successor was duly chosen — by Queen Victoria, in her capacity as Visitor of Christ Church. The new Dean was the Revd Henry Liddell (pronounced 'liddle'), Headmaster of Westminster School and half of the formidable team of Liddell and Scott, who produced the Greek–English Lexicon that the young Charles had purchased while at Rugby School; it is still in use by undergraduates today. The Liddells came with four children, one of whom, Alice (then aged three), would for ever be associated with the name of Lewis Carroll.

When the long summer vacation came, Dodgson delayed his customary return to Croft to make another visit to his Uncle Skeffington in London, where he spent a delightful week. Throughout his life Dodgson appreciated beauty in all its forms — whether in painting, poetry, mathematics, music or the theatre — and he cherished a memorable evening at the Princess's Theatre:

The Revd Henry Liddell, Dean of Christ Church

"Henry VIII.," the greatest theatrical treat I ever had or expect to have. I had no idea that anything so superb as the scenery and dresses was ever to be seen on the stage. Kean was magnificent as Cardinal Wolsey, Mrs. Kean a worthy successor to Mrs. Siddons in Queen Catherine, and all the accessories without exception were good — but oh, that exquisite vision of Queen Catherine's! I almost held my breath to watch: the illusion is perfect, and I felt as if in a dream all the time it lasted. It was like a delicious reverie . . . Never shall I forget that wonderful evening, that exquisite vision . . . I never enjoyed anything so much in my life before.

In later years he would make frequent excursions to London to see the latest plays and operettas — those of Gilbert and Sullivan being a particular favourite.

On his visits to Croft Rectory during vacations, he continued to produce family magazines for his younger brothers and sisters. *Mischmasch* contains some helpful 'Hints for Etiquette: or, Dining Out Made Easy':

V In proceeding to the dining-room, the gentleman gives one arm to the lady he escorts — it is unusual to offer both.

VIII The practice of taking soup with the next gentleman but one is now wisely discontinued.

IX To use a fork with your soup, intimating at the same time to your hostess that you are reserving the spoon for the beefsteaks, is a practice wholly exploded.

XI On meat being placed before you, there is no possible objection to your eating it, if so disposed.

XVII We do not recommend the practice of eating cheese with a knife and fork in one hand and a spoon and wine-glass in the other.

XXVI As a general rule, do not kick the shins of the opposite gentleman under the table, if personally unacquainted with him: your pleasantry is liable to be misunderstood.

Over the summer period, the new Dean appointed Dodgson to the Mathematical Lectureship which he so desired, to start at the beginning of 1856. Dodgson quickly made a resolution:

As I have learnt the Dean's intention of making me Mathematical Lecturer next term, I shall not go in for the Scholarship again.

Dodgson later recalled that

I had the Honourmen for the last two terms of 1855, but was not full Lecturer till Hilary 1856 . . . I gave my first Euclid Lecture in the Lecture-room on Monday January 28, 1856. It consisted of 12 men, of whom nine attended.

Charles also received a long letter from his father suggesting that he hold the Lectureship for a decade or so before settling down in a Christ Church parish. In such circumstances he might organize his finances as follows:

> I will just sketch for you a supposed case, applicable to your own circumstances, of a young man of twenty-three, making up his mind to work for ten years, and living to do it, on an Income enabling him to save £150 a year — supposing him to appropriate it thus:–

	£	s.	d.
Invested at 4 per cent	100	0	0
Life Insurance of £1,500	29	15	0
Books, beside those bought in ordinary course	20	5	0
	£150	0	0

> Suppose him at the end of ten years to get a Living enabling him to settle, what will be the result of his savings:–
>
> 1. A nest egg of £1,220 ready money, for furnishings and other expenses.
> 2. A sum of £1,500 secured at his death on payment of a *very much* smaller annual Premium than if he had begun to insure it.
> 3. A useful Library, worth more than £200, besides the books bought out of his current Income during the period.

One of the College traditions on the installation of a new Dean was for the Canons to appoint a 'Master of the House', an arrangement that gave the holder the privileges of a Master of Arts within the walls of the College. In October 1855 Dodgson was selected for this honour. He did not receive his official Master of Arts degree from the University until February 1857, after he had completed the required number of terms.

It had been an eventful year for the young Charles Dodgson. On 31 December 1855 he looked back on the past twelve months:

> I am sitting alone in my bedroom this last night of the old year, waiting for midnight. It has been the most eventful year of my life: I began it as a poor bachelor student, with no definite plans or expectations; I end it a master and tutor in Ch. Ch., with an income of more than £300 a year, and the course of mathematical tuition marked out by God's providence for at least some years to come. Great mercies, great failings, time lost, talents misapplied — such has been the past year.

A Spot of Schoolteaching

During the long summer vacation of 1855, Dodgson had taught a few classes at the new National School at Croft that his father had built on his land:

8 July: I took the first and second class of the Boy's School in the morning — we did part of the life of St. John, one of the "lessons" on Scripture lives. I liked my first attempt in teaching very much.

10 July: As there was nothing for me to do in the Boy's School, I took the second class in the girl's, and liked the experiment very much. The intelligence of the children seemed to vary inversely as their size. They were a little shy this first time, but answered well nevertheless.

16 July: All this week I took the first class of boys: besides this I teach James Coates Euclid and Algebra on Tuesday, Thursday and Friday evenings, and read Latin with Mr. Hobson on Wednesday evening and Saturday morning — so I get tolerable practice in teaching.

A second opportunity for schoolteaching arose in Oxford the following winter, when he tried his hand at some mathematics teaching at St Aldate's School, directly across the road from Christ Church:

29 January: I gave the first lesson there today, to a class of 8 boys, and found it more pleasant than I expected. The contrast is very striking between town and country boys: here they are sharp, boisterous, and in the highest spirits, the difficulty of teaching being not to get an answer, but to prevent all answering at once.

Dodgson varied his lessons with stories and puzzles, and he may have been the first to use recreational topics as a vehicle for conveying more serious mathematical ideas — the three tricks mentioned in the following diary entries are described in detail below. At first all went well:

5 February: Varied the lesson at the school with a story, introducing a number of sums to be worked out. I also worked for them the puzzle of writing the answer to an addition sum, when only one of the five rows has been written: this, and the trick of counting alternately

Charles Dodgson at Croft Rectory

up to 100, neither putting on more than 10 to the number last named, astonished them not a little.

But the class's enthusiasm soon waned:

8 February: The school class noisy and inattentive, the novelty of the thing is wearing off, and I find them rather unmanageable. Showed them the "9" trick of striking out a figure, after subtracting a number from its reverse. Was a good deal tired with the six hours' consecutive lecturing.

The three tricks mentioned above are ones that Dodgson enjoyed showing to young people. We present them here; explanations of how they work are given in the Notes at the end of the book.

The addition sum

In this puzzle you give me any four-digit number, say 2879; I then write down another number (22877) on a sheet of paper and hide it away. We now take it in turns to propose four more numbers: for example, you choose 4685, I choose 5314, you choose 7062 and I choose 2937. If you now add these five numbers together, you will find that their total is indeed 22877, the number I predicted. How did I arrange this?

Counting alternately up to 100

In this trick we start with the number 1 and then take it in turns to add a new number, never exceeding 10; the person that reaches 100 is the winner. For example, if to the 1 you first add 4 (giving 5), I'll then add 7 to give 12; you might then add 8 to give 20, and I'll then add 3 to give 23. Continuing in this way we might get the sequence 29, 34, 43, 45, 48, 56, 57, 67, 75, 78, 86, 89, 92, 100, and I win! How can I ensure that I am *always* the one to reach 100?

The '9' trick

Here I ask you to choose any number, reverse it, and then subtract the smaller number from the larger. From the answer you should then select any digit other than 0, remove it, and tell me the sum of the remaining digits: I will then tell you which number you removed. For example, if you choose 6173, and subtract its reverse, 3716, you get 2457; if you then choose to remove 5, and give me the sum of the remaining numbers (2 + 4 + 7 = 13), I can then tell you immediately that it was 5 that you removed. How do I know?

The situation in the classroom continued to deteriorate, and before long Dodgson felt that he had had enough:

> 26 *February*: Class again noisy and inattentive, it is very disheartening, and I almost think I had better give up teaching there for the present.
> 29 *February*: Left word at the school that I shall not be able to come again for the present. I doubt if I shall try again next term: the good done does not seem worth the trouble.

Dodgson as a Teacher

How good a teacher was Dodgson? Opinions seem to differ widely. One former undergraduate was complimentary:

> I was up at Christ Church as an undergraduate early in the eighties, he being my mathematical tutor and certainly his methods of explaining the elements of Euclid gave me the impression of being extremely lucid, so that the least intelligent of us could grasp at any rate 'the Pons Asinorum' [a result on isosceles triangles].

But one was less impressed:

> Very few, if any, of my contemporaries survive to confirm my impression of the singularly dry and perfunctory manner in which he imparted instruction to us, never betraying the slightest personal interest in matters that were of deep concern to us.

Another recalled that Dodgson

> was in the habit of punishing absentees from his lectures by giving them lines, which were discharged, for a small fee, by an old man named Boddington who dwelt in Oriel Lane.

Yet another remembrance was more cryptic:

> glancing at a problem in Euclid which I had written out, he placed his finger on an omission. "I deny you the right to assert that." I supplied what was wanting. "Why did you not say so before? What is a corollary?" Silence. "Do you ever play billiards?" "Sometimes." "If you attempted a cannon, missed, and holed your own and the red

ball, what would you call it?" "A fluke." "Exactly. A corollary is a fluke in Euclid. Good morning."

It has even been claimed that some undergraduates made an official complaint to the College authorities about Dodgson's teaching, but there seems to be no evidence to support this. The liveliness and effectiveness of his teaching probably varied from pupil to pupil and from topic to topic.

Whatever the quality of Dodgson's teaching, towards the end of 1856 he became overwhelmed by his teaching commitments. His lectures occupied seven hours a day, and he had to cover a wide range of topics, leaving him inadequate spare time for preparation. He felt himself daily becoming more and more unfit for the position he held.

12 November: I am becoming embarrassed by the duties of the Lectureship, and must take a quiet review of my position, to see what can be done . . . I have five pupils, whose lectures need preparing for, namely

Blackmore in for a First at Easter, doing end of Differential Calculus (*new to me*), and to begin Integral Calculus soon.

Rattle in for a First in Mods this time, needs special problems etc. and very probably high Diff: Cal:, a little Int: Cal: and Spherical Trig.

Blore in for a Second, easier problems etc.

Bradshaw in this time next year, reading the circle in *Salmon*, and is already in work new to me.

Harrison in for the Junior Scholarship this term, we are beginning *Salmon*, so that his case is included in Bradshaw's, and he is reading with Price as well, which makes his case easier . . .

Something must be done, and done *at once*, or I shall break down altogether.

Teaching for Responsions was also proving to be unrewarding:

26 November: I examined six or eight men today who are going in for Little-Go, and hardly one is really fit to go in. It is thankless uphill work, goading unwilling men to learning they have no taste for, to the inevitable neglect of others who really want to get on.

THE

T R A I N:

A First-Class Magazine.

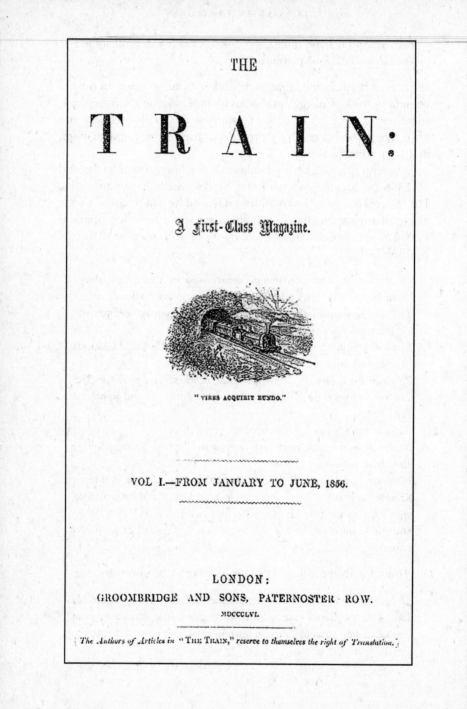

"VIRES ACQUIRIT EUNDO."

VOL I.—FROM JANUARY TO JUNE, 1856.

LONDON:

GROOMBRIDGE AND SONS, PATERNOSTER ROW.

MDCCCLVI.

The Authors of Articles in "THE TRAIN," reserve to themselves the right of Translation.

Fortunately, the situation gradually resolved itself as Dodgson gained confidence and experience, and he continued to hold the lectureship for a further twenty-five years.

Poems and Photographs

Around this time Dodgson began to publish his writings and verses. In mid-1855 he had made contact with Edmund Yates, editor of the *Comic Times* and its successor, *The Train: A First-Class Magazine*, and offered him poems, parodies and short stories, several of which duly appeared.

Yates suggested that he adopt a pseudonym for his comic writings so as to distinguish them from academic publications. Dodgson initially proposed *Dares* (short for Daresbury), which was rejected, but shortly afterwards he offered four other suggestions:

Wrote to Mr. Yates, sending him a choice of names,
1. *Edgar Cuthwellis* (made by transposition out of "Charles Lutwidge"),
2. *Edgar U. C. Westhill* (ditto),
3. *Louis Carroll* (derived from Lutwidge = Ludovic = Louis, and Charles),
4. *Lewis Carroll* (ditto).

As we all know, the last of these was selected. He used it when writing books for a general audience, such as his *Alice* books and his books on logic.

One of Dodgson's publications in *The Train* was 'Hiawatha's Photographing', an amusing parody of Henry Wadsworth Longfellow's lengthy poem *The Song of Hiawatha*, with its repetitive metre derived from a Finnish epic poem:

> By the shore of Gitche Gumee,
> By the shining Big-Sea-Water,
> At the doorway of his wigwam,
> In the pleasant summer morning,
> Hiawatha stood and waited.

Longfellow's poem was written in 1855 and chronicled the life of Hiawatha, the wise man and orator with supernatural gifts who was sent by the Great Peacemaker to guide the Indian nations. Two years later, Dodgson described Hiawatha's unsuccessful attempts to photograph the members of a highly dysfunctional family:

From his shoulder Hiawatha
Took the camera of rosewood —
Made of sliding, folding rosewood —
Neatly put it all together.
In its case it lay compactly,
Folded into nearly nothing;
But he opened out the hinges,
Pushed and pulled the joints and hinges,
Till it looked all squares and oblongs,
Like a complicated figure
In the second book of Euclid.

Amateur photography had become the popular craze of the 1850s. In September 1855, Dodgson's Uncle Skeffington, always one for the latest gadgets, visited the family at Croft with his camera, and Dodgson joined him in some photographic excursions.

He became addicted, and remained so for the next twenty-five years. At Christ Church he befriended Reginald Southey, an undergraduate studying medicine who was an accomplished photographer, and on 18 March 1856 they travelled to London, where Dodgson purchased a fine Ottewill rosewood box camera and lens for the substantial sum of £15.

The wet collodion process, which Dodgson used, required great skill and perseverance, and failures were frequent. It involved posing the sitter, going into a darkroom, covering a glass plate with collodion (a glutinous substance), sensitizing the plate with a silver nitrate solution, returning to the sitter, removing the lens while the sitter remained motionless for a short time, returning to the darkroom with the exposed plate, developing it in a carefully prepared chemical solution, fixing it in another solution, heating and varnishing the plate to create a negative, and finally producing the desired positive images. In his poem, Dodgson details how Hiawatha used the hazardous chemicals:

> First, a piece of glass he coated
> With collodion, and plunged it
> In a bath of lunar caustic
> Carefully dissolved in water —
> There he left it several minutes.
>
> Secondly, my Hiawatha
> Made with cunning hand a mixture
> Of the acid pyrro-gallic,
> And of glacial-acetic,
> And of alcohol and water —
> This developed all the picture.
>
> Finally, he fixed each picture
> With a saturate solution
> Which was made of hyposulphite,
> Which, again, was made of soda.

It took Dodgson many months of experimentation to become adept — but adept he certainly became, and if he were not known

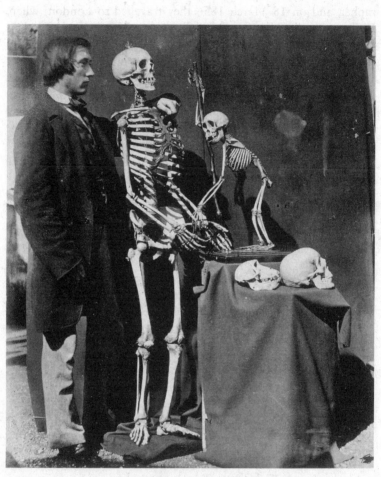

Reginald Southey and friends
(photographed by Charles Dodgson, 1857)

for his *Alice* books he would now be remembered as one of the most important photographers of the nineteenth century. He was one of the earliest to consider photography as an art form, rather than simply as a means of recording images. By carefully positioning his subjects, and with appropriate choices of costume and props, he obtained many memorable results; his depiction of Southey with skeletons is a good example.

In total, Dodgson took around three thousand photographs, most of which are portraits of individuals or small groups. Some of them are of his Oxford contemporaries and give us a valuable insight into university life over a period of twenty years. Others are of family members, such as his sensitive portrayals of his brother Edwin (at the age of thirteen) and sister Maggie (aged eighteen), and two of his maternal aunts playing chess at Croft Rectory.

Although he has often been described as shy and laconic, Dodgson was a sociable character whose photographic talents opened doors and enabled him to meet many distinguished personages of the time. By his early thirties he had already photographed such luminaries as the poet Alfred Tennyson, the scientist Michael Faraday, the actress Ellen Terry, and the artists William Holman Hunt, John Millais, John Ruskin and Dante Gabriel Rossetti. Several of these became personal friends.

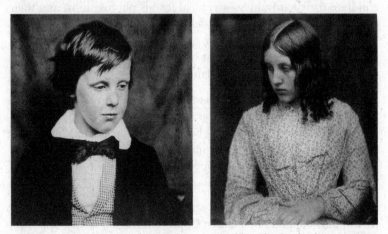

Edwin and Margaret Dodgson (photographed by Charles Dodgson, 1859)

Margaret Anne and Henrietta May Lutwidge
(photographed by Charles Dodgson, 1859)

But Dodgson is primarily remembered for his artistic images of children: their naturalness was ideal for his photographs, and their curiosity was enchanting. He delighted in showing them how he mixed his chemicals, providing exotic costumes for them to wear, and sharing in their excitement as the images gradually appeared on the plates. As one of them later recalled:

A visit to Mr. Dodgson's rooms to be photographed was always full of surprises. Although he had quaint fancies in the way he dressed his little sitters, he never could bear a dressed-up child. A 'natural child' with ruffled untidy hair suited him far better, and he would place her in some ordinary position of daily life, such as sleeping, or reading, and so produce charming pictures.

An excellent example is his study of Xie (pronounced 'exie') Kitchin, the daughter of a Christ Church colleague, shown opposite.

Xie Kitchin — 'Rosy dreams and slumbers light'
(photographed by Charles Dodgson, 1873)

Dodgson's contribution to the subject has been described by Helmut Gernsheim, the renowned historian of photography, in the following terms:

his photographic achievements are truly astonishing: he must not only rank as a pioneer of British amateur photography, but I would also unhesitatingly acclaim him as the most outstanding photographer of children in the nineteenth century.

Ciphers

Throughout his life, Dodgson enjoyed word games. In 1856 he noted that

I am thinking of writing an article on "Cipher" for the *Train*, but must first consult Mr. Yates as to whether the subject will be admissible.

But nothing appeared, and it was to be a further two years before he returned to the subject.

In February 1858 he invented a *key-vowel cipher*. In such a cipher the sender uses a keyword to encode a given message, replacing each of its letters by another one. The receiver, who also knows the keyword, then reverses the process to recover the original message.

Dodgson was clearly dissatisfied with his first cipher, for two days later he recorded that he had devised another one, far better than the last:

> It has these advantages.
> (1) The system is easily carried in the head.
> (2) The key-word is the only thing necessarily kept secret.
> (3) Even one knowing the system cannot possibly read the cipher without knowing the key-word.
> (4) Even with the English to the cipher given, it is impossible to discover the key-word.

This new one was his *matrix cipher*, which is based on the following grid:

A	F	L	Q	W
B	G	M	R	X
C	H	N	S	Y
D	IJ	O	T	Z
E	K	P	UV	*

Here *I* and *J* occupy the same position, as do *U* and *V*, and the last square is empty. We imagine that the first column follows the last one (so *B* follows *X*), and likewise the top row follows the bottom one (so *L* follows *P*).

Following Dodgson, let us suppose that the keyword is *GROUND*, known only to the sender and the receiver, and that the first word of our message is *SEND*:

- to encode the letter *S* we go from *G* (the first letter of the keyword) to *S*: this is 2 places to the right and 1 place down, and we encode *S* as *21*;
- to encode the letter *E* we go from *R* (the next letter of the keyword) to *E*: this is 2 places to the right and 3 places down, and we encode *E* as *23*;

- to encode the letter *N* we go from *O* to *N*: this is 0 places to the right and 4 places down, and we encode *N* as *04*;
- to send the letter *D* we go from *U* to *D*: this is 2 places to the right and 4 places down, and we encode *D* as *24*.

The message is now encoded as *21–23–04–24*, and may be sent in this form to the receiver.

To recover the original message, the receiver reverses the process:

- to decode the symbol *21*, start from *G* (the first letter of the keyword) and go 2 places to the right and 1 place down: this gives *S*;
- to decode the symbol *23*, start from *R* (the next letter of the keyword) and go 2 places to the right and 3 places down: this gives *E*;
- to decode the symbol *04*, start from *O* and go 0 places to the right and 4 places down: this gives *N*;
- to decode the symbol *24*, start from *U* and go 2 places to the right and 4 places down: this gives *D*.

The message is therefore *SEND*.

If you would like to try out this code, here is a Carrollian exhortation, encoded by means of the matrix cipher with the keyword *ALICE*, repeated several times (the decoded message can be found in the Notes at the end of the book):

01–34–32–03–32–04–13–04–02–14–00–31–
43–02–32–40–03–44–12–22–42–12–10–20

Dodgson's interest in ciphers was timely. The first electric telegraph, from Paddington to Slough on the Great Western Railway, had opened in 1843, and for the next two decades ciphers were increasingly employed, particularly in military situations, to protect messages from unwarranted interception. Dodgson offered his ciphers to the Admiralty, but seems to have received no response; he also used them in his letters to children.

Around 1868, he returned to the subject of ciphers and invented two further ones — his *alphabet cipher* and *telegraph cipher*. The former, which we describe here, is his rediscovery of the sixteenth-century *Vigenère code*, and is based on a 26 × 26 grid, each row of which provides a separate method of encoding; for his first cipher of 1858, the grid had used only five rows, corresponding to the vowels.

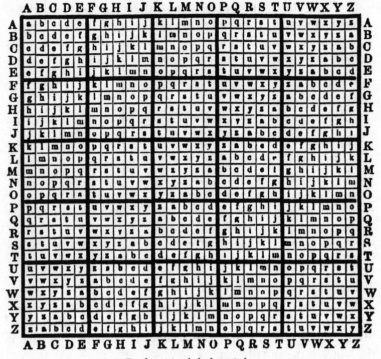

Dodgson's alphabet cipher

Originally printed on a piece of card measuring about 7 × 5 inches, the alphabet cipher carried instructions on the back.

As before, we need a keyword, known only to the sender and the receiver; following Dodgson, let us take the word *VIGILANCE*. To send a message such as

MEET ME ON TUESDAY EVENING AT SEVEN

we write out the following table:

V I G I L A N C E V I G I L A N C E V I G I L A N C E V I
M E E T M E O N T U E S D A Y E V E N I N G A T S E V E N

We then proceed column by column, as follows:

- to encode the letter *M*, write down the letter in row *V* and column *M*: this is *h*;

78

- to encode the first *E*, write down the letter in row *I* and column *E*: this is *m*;
- to encode the second *E*, write down the letter in row *G* and column *E*: this is *k*;
- to encode the letter *T*, write down the letter in row *I* and column *T*: this is *b*.

Continuing in this way, we eventually obtain the encoded message as

h m k b x e b p x p m y l l y r x l l q t o l t f g z z v

To decode this message, the receiver reverses the process, still using the keyword:

- to decode the letter *h*, look for it in row *V*: it is in column *M*;
- to decode the letter *m*, look for it in row *I*: it is in column *E*;
- to decode the letter *k*, look for it in row *G*: it is in column *E*;
- to decode the letter *b*, look for it in row *I*: it is in column *T*.

Continuing in this way, we recapture the original message.

We conclude this chapter with another Carrollian quotation for you to try, encoded by means of the alphabet cipher with the keyword *ALICE*. The decoded message can be found in the Notes at the end of the book:

t h i u f r t t n m g l v f x h p a n m t s g v s v p a

Charles Dodgson polishes his camera lens

Fit the Fourth
. . . in the Second Book of Euclid

In 'Hiawatha's Photographing', Hiawatha's box camera is described as looking 'all squares and oblongs, like a complicated figure in the second book of Euclid'. But who was Euclid, and why was he so important to Charles Dodgson?

From his earliest years, Dodgson had been enthused by geometry. We have already seen (in Fit the First) a construction that he described at the age of twelve, and later, while at Rugby School, he wrote to his sister Elizabeth suggesting a geometrical name for one of her new pets:

> I am glad to hear of the 6 rabbits. For the new name after some consideration I recommend Parellelopipedon — It is a nice easy one to remember, and the rabbit will soon learn it.

Parallelopipedon (as he should have written it) is the Greek word for a parallelepiped, a skewed box bounded by parallelograms.

The hues of life are dull and gray,
The sweets of life insipid,
When *thou*, my charmer, art away —
Old Brick, or rather let me say,
Old Parallelepiped!

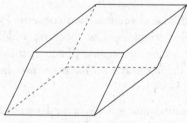

Parallelepiped

In his books for children, such as *Sylvie and Bruno*, Dodgson sometimes resorted to geometrical word-play:

> **Professor:** (*drawing a long line on the black board, and marking the letters "A", "B", at the two ends, and "C" in the middle*) If *AB* were to be divided into two parts at *C* —
> **Bruno:** It would be drownded.
> **Professor:** *What* would be drownded?
> **Bruno:** Why the bumble-bee, of course! And the two bits would sink down in the sea!

With many cheerful facts about
the square of the hypotenuse.
W. S. Gilbert

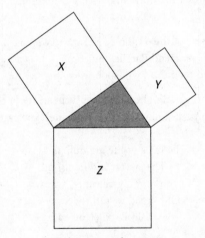

Pythagoras's theorem

There is also a whimsical account that concerns Pythagoras's theorem on right-angled triangles. The theorem, which states that *the area of the square on the hypotenuse* (the longest side) *is equal to the sum of the areas of the squares on the other two sides* (Z = X + Y), was, according to Dodgson,

> as dazzlingly beautiful now as it was in the day when Pythagoras first discovered it, and celebrated its advent, it is said, by sacrificing a hecatomb of oxen — a method of doing honour to Science that has

always seemed to me *slightly* exaggerated and uncalled-for. One can imagine oneself, even in these degenerate days, marking the epoch of some brilliant scientific discovery by inviting a convivial friend or two, to join one in a beefsteak and a bottle of wine. But a *hecatomb* of oxen! It would produce a quite inconvenient supply of beef.

However, the Greek author whom Dodgson most admired was not Pythagoras, but Euclid.

Here's Looking at Euclid

Euclid of Alexandria (now in Egypt) lived around 300 BC and wrote the most widely used mathematical text of all time, the *Elements*. For over two thousand years — from the academies of ancient Greece and the universities of medieval Europe to the private schools of Victorian England — this text was used to teach geometry and train the mind. The *Elements* is believed to be the most printed book of all time, after the Bible: during the nineteenth century, over two hundred editions were published in England alone, with one popular version selling over half a million copies.

But these many editions differed greatly in style and content. Over the Easter vacation of 1855, Charles Dodgson started teaching geometry to Louisa, the most mathematically gifted of his sisters:

> Went into Darlington — bought at Swale's, *Chamber's Euclid* for Louisa. I had to scratch out a good deal he had interpolated, (e.g. definitions of words of his own) and put some he had left out. An author has no right to *mangle* the original writer whom he employs: all additional matter should be carefully distinguished from the genuine text. N.B. Pott's [*sic*] *Euclid* is the only edition worth getting — both Capell and Chamber's are mangled editions.

Robert Potts's edition, Dodgson's preferred choice, was *The School Edition, Euclid's Elements of Geometry, the first six books, chiefly from the text of Dr. Simson, with Explanatory Notes.*

Euclid's *Elements* consists of thirteen 'Books', organized as follows:

Books I–IV introduce the basic geometry of the plane: points, lines, angles, triangles, parallel lines, rectangles and circles. A typical result on triangles is the *pons asinorum*, or asses' bridge (Book I,

Proposition 5), which states that *in any isosceles triangle* (a triangle with two equal sides) *two of the angles are equal*; in medieval European universities, if you understood this proposition then you could cross the asses' bridge and proceed to the delightful results beyond, such as Pythagoras's theorem (Book I, Proposition 47).

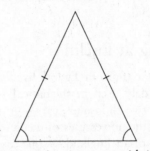

The pons asinorum, *or asses' bridge*

Books V and VI are concerned with similar figures — those of the same shape but not necessarily the same size; for example, the following triangles are similar:

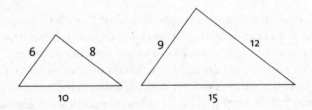

A typical deduction in *Book V*, expressed in terms of ratios of lengths, is that *if the ratio a : b equals the ratio c : d, then the ratio a : c equals the ratio b : d.* For example, knowing that 6 : 8 = 9 : 12, we can deduce that 6 : 9 = 8 : 12. These books also discussed 'commensurable' magnitudes — pairs of lengths whose ratio can be written in terms of whole numbers, such as 0.03 : 0.04 and $3\pi : 4\pi$, which are both in the ratio 3 : 4. Two lengths that are *incommensurable* are the side and diagonal of a square, because these lengths are in the ratio of 1 : ≠2, which cannot be written in terms of whole numbers.

The later books do not concern us directly. *Books VII–IX* deal with the arithmetic of odd and even numbers, squares and cubes, and prime numbers. *Book X* is a long and complicated account of commensurable numbers. *Books XI–XIII* present the geometry of three dimensions: solids, pyramids, spheres, cones and cylinders.

The *Elements* is organized in a hierarchical way, building on certain initial statements called *definitions*, *postulates* and *axioms*. Here are three definitions from Book I that you will meet again later:

- A plane angle *is the inclination of two straight lines to one another, which meet together, but which are not in the same direction.*
- *When a line, meeting another line, makes the angles on one side equal to that on the other, the angle on each side is called a* right angle.

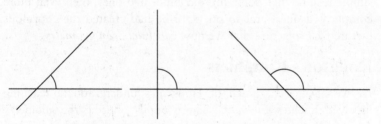

A plane angle, a right angle and an obtuse angle

- *An* obtuse angle *is one which is greater than a right angle.*

Dodgson described a postulate as 'something to be done, for which no proof is given'. Euclid listed five postulates, starting with:

- *Let it be granted, that a line may be drawn from any point to any other point.*
- *That a finite line may be produced (lengthened) to any extent.*

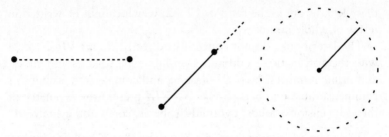

Drawing a line between points, producing a finite line and drawing a circle

- *That a circle may be drawn about any point, and at any distance from that point.*

He also described an axiom as 'something to be believed, for which no proof is given'. Euclid listed five axioms, starting with:

- *Things that are equal to the same thing are equal to each other.*

In modern notation, this says that if $A = X$ and $B = X$, then $A = B$.

From these humble beginnings Euclid first deduced some very simple results, then some more complicated ones, then even more complicated ones, and so on, until he had created the enormous hierarchical structure that we now call *Euclidean geometry*.

Dodgson's Pamphlets

As we have seen, much of Dodgson's undergraduate teaching involved Euclid's *Elements*. Those pupils offering the geometry paper in Responsions would need to know the material of Books I and II, while passmen taking the Euclid paper in Moderations would also study Book III, and classmen taking mathematics Finals would continue up to Book VI.

Although Dodgson greatly admired Euclid's *Elements*, he realized that it contained some gaps, inaccuracies and inconsistencies. In order to help his pupils overcome these, he produced a number of short mathematical pamphlets that clarified the text, suggested alternative approaches, and included exercises for his pupils to attempt. Over the years he became an enthusiastic producer of pamphlets, and published more than two hundred on a wide range of topics.

His first mathematical pamphlet was *Notes on the First Two*

Books of Euclid (1860), costing sixpence and designed to help those taking Responsions. These notes were subsequently expanded into book form, as *Euclid, Books I, II*, which appeared in an unpublished private version in 1875 and were published in 1882, eventually running to eight editions. Their aim was

> to show what Euclid's method really is in itself, when stripped of all accidental verbiage and repetition. With this object, I have held myself free to alter and abridge the language wherever it seemed desirable, so long as I made no real change in his methods of proof, or in his logical sequence.
>
> The result is that the text of this Edition is (as I have ascertained by counting the words) *less than five-sevenths* of that contained in the ordinary Editions.

Associated with his *Notes* was a useful pamphlet entitled *The Enunciations of Euclid I, II* (1863), in which he listed all the main definitions and propositions needed for Responsions; ten years later he expanded this to *The Enunciations of Euclid I–VI* for those taking Finals.

In 1868 he produced a pamphlet entitled *The Fifth Book of Euclid Treated Algebraically*, in which he explained each definition with examples and recast each proposition in a more accessible algebraic form. For simplicity he omitted the theory of incommensurable numbers as being inappropriate for undergraduates.

While these pamphlets and books were generally well received and widely used, not everyone was enthusiastic. In particular, Robert Potts, editor of Dodgson's preferred version of the *Elements*, had concerns about spoon-feeding the readers:

> I have had considerable experience in dealing with minds of low logical power, and have found that studies may be made so easy and mechanical as to render thought almost superfluous.

The Dynamics of a Parti-cle

Not all of Dodgson's geometrical endeavours were so serious. His witty pamphlet *The Dynamics of a Parti-cle* satirized the parliamentary election for the Oxford University seat in July 1865. In the Victorian era, Oxford University was represented by two

Members of Parliament in London. In 1865 the candidates for these seats were the sitting candidate William Ewart Gladstone, who had represented the University for eighteen years but who was too liberal for Dodgson; the more conservative Gathorne Gathorne-Hardy, who was Dodgson's preferred choice; and a third candidate, Sir William Heathcote.

Much of the pamphlet resembles a treatise on formal geometry, with some initial 'definitions' parodying the Euclidean ones we saw earlier:

- Plain anger *is the inclination of two voters to one another, who meet together, but whose views are not in the same direction.*
- *When a Proctor, meeting another Proctor, makes the votes on one side equal to those on the other, the feeling entertained by each side is called* right anger.
- Obtuse anger *is that which is greater than* right anger.

Dodgson then introduced his 'postulates', which again mimic Euclid's:

- *Let it be granted, that a speaker may digress from any one point to any other point.*
- *That a finite argument (i.e. one finished and disposed of) may be produced to any extent in subsequent debates.*
- *That a controversy may be raised about any question, and at any distance from that question.*

And so he continued for several pages, leading up to the following geometrical construction. It is designed

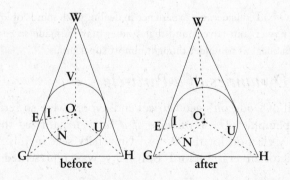

> To remove a given Tangent from a given Circle, and to bring another given Line into contact with it.

In order to do so, Dodgson assigned letters to the points of the diagram:

> Let UNIV be a Large Circle, whose centre is O (V being, of course, placed at the top), and let WGH be a triangle, two of whose sides, WEG and WH, are in contact with the circle, while GH (called "the base" by liberal mathematicians) is not in contact with it. It is required to destroy the contact of WEG, and to bring GH into contact instead.

Here, UNIV represents the University, O is Oxford, V is the Vice-Chancellor, and WEG, GH and WH are the three candidates; the object is to unseat Gladstone and replace him by Gathorne-Hardy. Before working through a pseudo-geometrical argument, Dodgson observes that

> When this is effected, it will be found most convenient to project WEG to infinity.

This indeed happened: Heathcote and Gathorne-Hardy were duly elected.

The Euclid Debate

As we have seen, the Victorian curriculum in most English private schools consisted mainly of the classical languages, together with some Divinity. For those schools that taught mathematics, Euclid's geometry was the standard fare, being regarded as the ideal vehicle for teaching young men how to reason and think logically. Based on 'absolutes', the study of geometry fitted well with the classical curriculum, thereby providing a suitable training for those expecting to go on to Oxford and Cambridge and then into the Church. The *Elements* thus became an important constituent of examination syllabuses, being required also for entrance to the civil service and the army.

The benefits of geometry were extolled by the philosopher William Whewell:

> There is no [other] study by which the Reason can be so exactly and rigorously exercised. In learning Geometry the student is rendered

familiar with the most perfect examples of strict inference . . . He is accustomed to a chain of deduction in which each link hangs from the preceding, yet without any insecurity in the whole: to an ascent, beginning from the solid ground, in which each step, as soon as it is made, is a foundation for the further ascent, no less solid than the first self-evident truths.

We require our present Mathematical studies not as an instrument (for the solution of today's mathematical problems) but as an exercise of the intellectual powers; that is, not for their results, but for the intellectual habits which they generate that such studies are pursued.

Others, however, were opposed to the dry and over-formal approach of Euclid, regarding such a strictly logical approach as obscure, unsuitable for beginners, and artificial in its insistence on a minimal set of axioms. Another objection was that the formal study of Euclid failed to encourage independent thinking, requiring too much rote learning, often with no understanding; indeed, the story was told of an Oxford examination candidate who reproduced a proof from Euclid perfectly, except that in his diagram he drew all the triangles as circles.

It was a time of change. A growing middle class was demanding a more practical approach to mathematics, and the traditional classical education was becoming increasingly irrelevant. In his 1869 Presidential address to the British Association for the Advancement of Science, the mathematician James Joseph Sylvester was forthright in his condemnation of the old ways:

> The early study of Euclid made me a hater of geometry, which I hope may plead my excuse if I have shocked the opinions of any in this room (and I know there are some who rank Euclid as second in sacredness to the Bible alone, and as one of the advanced outposts of the British constitution) . . . I think that study and mathematical culture should go on hand in hand together, and that they would greatly influence each other for their mutual good. I should rejoice to see . . . Euclid honourably shelved or buried 'deeper than e'er plummet sounded' out of the schoolboy's reach.

Throughout the 1860s, the feeling grew that examinations

should no longer be based on a single book. Several texts were proposed as alternatives to the *Elements* — at first a trickle, then a flood. A Schools' Inquiry Commission was set up, which, in the forthright words of the mathematician Augustus De Morgan,

> raised the question whether Euclid be, as many suppose, the best elementary treatise on geometry, or whether it be a mockery, delusion, snare, hindrance, pitfall, shoal, shallow, and snake in the grass.

Following Sylvester's lecture to the British Association, an Anti-Euclid Association was formed, and January 1871 saw the foundation of an Association for the Improvement of Geometrical Teaching, which took on the task of producing new geometry syllabuses and texts.

Euclid and his Modern Rivals

Charles Dodgson, the outspoken advocate for Euclid's *Elements*, was bitterly opposed to the new Association and its aims. In 1879 he wrote a remarkable work, *Euclid and his Modern Rivals*, in which he skilfully compared the *Elements*, favourably in each case, with thirteen well-known rival texts.

Dodgson introduced his book, which was 'Dedicated to the memory of Euclid', as follows:

> The object of this little book is to furnish evidence, first, that it is essential, for the purpose of teaching or examining in elementary Geometry, to employ one textbook only; secondly, that there are strong *a priori* reasons for retaining, in all its main features, and specially in its sequence and numbering of Propositions and in its treatment of Parallels, the Manual of Euclid; and thirdly, that no sufficient reasons have yet been shown for abandoning it in favour of any one of the modern Manuals which have been offered as substitutes.

He referred particularly to the numbering of familiar results:

> The Propositions have been known by those numbers for two thousand years; they have been referred to, probably, by hundreds of writers ... and some of them, I. 5 and I. 47 for instance — 'the Asses' Bridge' and 'the Windmill' [Pythagoras's theorem] — are now

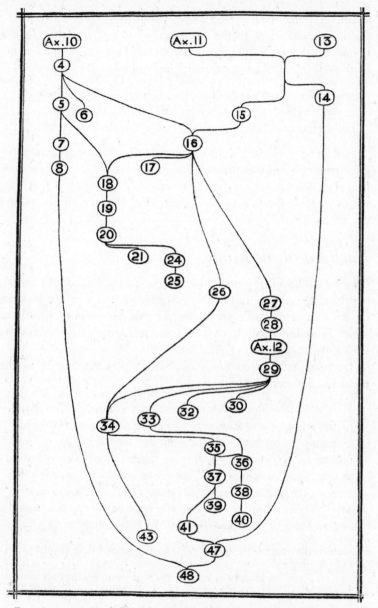

Frontispiece to Euclid and his Modern Rivals: *Dodgson's hierarchical arrangement of the Propositions in Euclid, Book I*

historical characters, and their nicknames are 'familiar as household words.'

Attempting to reach a wider audience, Dodgson cast his book as a play in four acts:

It is presented in dramatic form, partly because it seemed a better way of exhibiting in alternation the arguments on the two sides of the question; partly that I might feel myself at liberty to treat it in a rather lighter style than would have suited an essay, and thus to make it a little less tedious and a little more acceptable to unscientific readers.

There are four characters: Minos and Radamanthus (two of the three judges in Hades, appearing here as hassled Oxford examiners), Herr Niemand (the phantasm of a German professor who 'has read all books, and is ready to defend any thesis, true or untrue') and the ghost of Euclid himself.

The curtain rises on a College study at midnight, where Minos is wearily working his way through an enormous pile of examination scripts:

So, my friend! That's the way you prove I. 19, is it? Assuming I. 20? Cool, refreshingly cool! But stop a bit! Perhaps he doesn't 'declare to win' on Euclid. Let's see. Ah, just so! 'Legendre,' of course! Well, I suppose I must give him full marks for it: what's the question worth? — Wait a bit, though! Where's his paper of yesterday? I've a very decided impression he was all for 'Euclid' then: and I know the paper had I. 20 in it.

Ah, here it is! 'I think we do know the sweet Roman hand.' Here's the Proposition as large as life, and proved by I. 19. 'Now infidel, I have thee on the hip!' You shall have such a sweet thing to do in *vivâ-voce*, my very dear friend! You shall have the two Propositions together, and take them in any order you like. It's my profoundest conviction that you don't know how to prove either of them without the other.

Minos eventually falls asleep over his scripts, but is soon awakened by the ghost of Euclid, who invites him to compare the *Elements* with its Modern Rivals. To this end, Euclid summons up his friend Herr Niemand, who forcibly argues the case for each rival text. Minos carefully exposes their faults, in each case preferring Euclid's constructions, demonstrations, style, or treatment of the material. In a memorable stage direction, the syllabus of the Association for the Improvement of Geometrical Teaching comes in for particular ridicule:

> *Enter a phantasmic procession, grouped about a banner, on which is emblazoned in letters of gold the title* 'ASSOCIATION FOR THE IMPROVEMENT OF THINGS IN GENERAL.' *Foremost in the line marches* NERO, *carrying his unfinished 'Scheme for lighting and warming Rome'; while among the crowd which follow him may be noticed —* GUY FAWKES, *President of the 'Association for raising the position of Members of Parliament'.*

After demolishing each rival book in turn, Minos is approached by Euclid, together with the phantasms of mathematicians such as Archimedes, Pythagoras, Aristotle and Plato, who have come to see fair play:

> **Euclid:** Are all gone?
> **Minos:**   'Be cheerful, sir:
>   Our revels now are ended: these our actors,
>   As I foretold you, were all spirits, and
>   Are melted into air, into thin air!'
> **Euclid:** Good. Let us to business. And first, have you found any method of treating Parallels to supersede mine?
> **Minos:** No! A thousand times, no! The infinitesimal method, so gracefully employed by M. Legendre, is unsuited to beginners: the method by transversals, and the method by revolving Lines, have not yet been offered in a logical form: the 'equidistant' method is too cumbrous: and as for the 'method of direction', it is simply a rope of sand — it breaks to pieces wherever you touch it!
> **Euclid:** We may take it as a settled thing, then, that you have found no sufficient cause for abandoning either my sequence of Propositions or their numbering.

Before disappearing, Euclid makes his farewell speech:

'The cock doth craw, the day doth daw,' and all respectable ghosts ought to be going home. Let me carry with me the hope that I have convinced you of the importance, if not the necessity, of retaining my order and numbering, and my method of treating straight Lines, angles, right angles, and (most especially) Parallels.

The book was a tour de force, exhibiting Dodgson's intimate knowledge and deep understanding of Euclidean geometry. But ultimately it failed to achieve its aim. In 1888, Oxford and Cambridge, fearing that abandoning Euclid would reduce the examination systems to chaos, reluctantly agreed to accept proofs other than Euclid's, provided they did not violate Euclid's ordering of the propositions. By 1903 this restriction too had disappeared, when they agreed to accept any systematic treatment. A formal approach to geometry continued to be taught in many high schools until about the 1960s, when it was quietly dropped. Today, few schoolchildren are aware of the Euclidean approach to geometry.

Dodgson's Hexagon

One of the difficulties with the *Elements* was Euclid's fifth postulate (known to Victorians as his 12th axiom). The other four postulates are short and simple, but this one seems more complicated:

- *If a straight line falling on two straight lines makes the interior angles on the same side less than two right angles, then the two straight lines, if produced indefinitely, meet on that side on which the angles are less than the two right angles.*

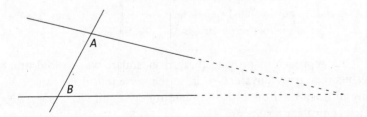

In other words, if, as illustrated here, angle *A* plus angle *B* is less than 180 degrees, then the two lines must eventually meet:

In his amusing introduction to *The Dynamics of a Parti-cle*, Dodgson referred to the fifth postulate as 'a striking illustration of the advantage of introducing the human element into the hitherto barren region of Mathematics':

> It was a lovely Autumn evening, and the glorious effects of chromatic aberration were beginning to show themselves in the atmosphere as the earth revolved away from the great western luminary, when two lines might have been observed wending their weary way across a plane superficies. The elder of the two had by long practice acquired the art, so painful to young and impulsive loci, of lying evenly between his extreme points; but the younger, in her girlish impetuosity, was ever longing to diverge and become a hyperbola or some such romantic and boundless curve. They had lived and loved: fate and the intervening superficies had hitherto kept them asunder, but this was no longer to be: *a line had intersected them, making the two interior angles together less than two right angles.* It was a moment never to be forgotten, and, as they journeyed on, a whisper thrilled along the superficies in isochronous waves of sound, "Yes! We shall at length meet if continually produced!"

For two thousand years, generations of mathematicians tried to deduce the fifth postulate from the others. In *Euclid and his Modern Rivals*, Minos and Euclid discuss the failure to do so:

> **Minos:** An absolute *proof* of it, from first principles, would be received, I can assure you, with absolute *rapture*, being an *ignis fatuus* [a delusive hope] that mathematicians have been chasing from your age down to our own.
>
> **Euclid:** I know it. But I cannot help you. Some mysterious flaw lies at the root of the subject.

One approach to proving the fifth postulate was to find other results that are 'equivalent' to it, in the sense that if we can prove any of them then the fifth postulate follows, and vice versa. Two such equivalent results are:

- The sum of the angles in any triangle is 180 degrees.
- Given any line *L* and any point *P* that does not lie on this line, there is exactly one line, parallel to *L*, that passes through *P*.

Strand Book Store

828 Broadway
New York, New York 10003
212.473.1452
strandbooks.com
strand@strandbooks.com

Date: 09-20-2011
Sale: 864696
Time: 05:34 PM

Lewis Carroll in Numberland: His Fantastical Mathe
0393304523   Item Price: $6.95 $6.95
Concise American Heritage Spanish Dictionary
0618117695   Item Price: $6.95 $6.95

2 Items   Subtotal: $13.90
                    Sales Tax (8.875 %): $1.23
                              Total: $15.13

        Credit (MasterCard) Payment: $15.13
Card Number ***********4059 XX/XX
ADAM M JACKREL
Approval # 42507313

                    Amount Tendered: $15.13
                       Change Due: $0.00

Cust Sign: _____

Manager Sign: _____

Printed by: aya   Register: CashDrawer4

        September 20, 2011 05:34 PM

            Thank you for shopping
                  at Strand
    Hardcovers/Merchandise are returnable
    within 3 days for a refund, Paperbacks
      & Clearance items are Final Sale

    Visit strandbooks.com for upcoming events.

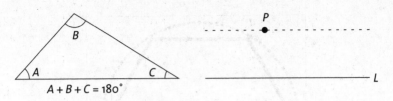

$$A + B + C = 180°$$

These results are pictured opposite. Because of the latter one, the fifth postulate became widely known as the *parallel postulate*.

Dodgson tried very hard to prove these results, but without success:

> Like the goblin 'Puck,' it has led me "up and down, up and down," through many a wakeful night: but always, just as I thought I had it, some unforeseen fallacy was sure to trip me up, and the tricksy sprite would "leap out, laughing ho, ho, ho!"

In fact, it is *not* possible to deduce these results from the other postulates. This was demonstrated around 1830 by Nikolai Lobachevsky in Russia and János Bolyai in Hungary, who independently constructed a strange type of geometry in which the other four postulates hold but the fifth does not.

Dodgson was aware of such 'non-Euclidean geometries', but rejected them as irrelevant to the geometrical world in which we live. In 1888 he produced a volume entitled *Curiosa Mathematica*, Part I: *A New Theory of Parallels*, in which he replaced the fifth postulate by a seemingly more 'obvious' one, pictured in the book's frontispiece (see overleaf). It asserts that *the area inside the hexagon is larger than the area of any one of the six pieces that lie between the hexagon and the circle*; he later replaced the hexagon by a square. By assuming the truth of this self-evident postulate, he was able to prove the fifth postulate and all the other results equivalent to it.

Squaring the Circle

> It is easier to square the circle than to get round a mathematician.
>
> *Augustus De Morgan*

Euclid's *Elements* contains a number of constructions for 'squaring'

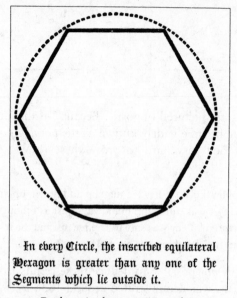

In every Circle, the inscribed equilateral Hexagon is greater than any one of the Segments which lie outside it.

Dodgson's alternative postulate

various shapes — that is, for constructing a square with the same area as the given shape, using only an unmarked ruler (for drawing straight lines, with no measuring permitted) and a pair of com-

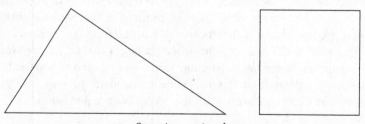

Squaring a triangle

passes (for drawing circles). In particular, Euclid shows how to square any given triangle, quadrilateral or pentagon.

One of the classical mathematical problems of ancient Greece was to find a corresponding construction for 'squaring the circle'.

Over the course of two millennia, many people believed that they had discovered such a construction, but all were wrong, and in popular culture the phrase 'squaring the circle' came to mean attempting the impossible. But it was not until 1882 that a German mathematician, Ferdinand Lindemann, proved once and for all that the task really is impossible.

The problem of squaring the circle is closely related to the mathematical properties of the number $\pi = 3.1415926535897\ldots$, the ratio of the circumference of a circle to its diameter. This number cannot be written down exactly — it 'goes on for ever' — but you can remember its first few digits by counting the number of letters in each word of the following sentence:

May I have a large container of coffee?
3 1 4   1   5   9   2   6

or, to go a little further,

How I need a drink, alcoholic of course, after all those chapters involving Dodgson.
3 1 4 1 5   9   2   6   5   3   5   8   9   7

For many years, Dodgson was plagued by cranks who sent him supposed constructions for squaring the circle, or 'proofs' that π has an exact value different from the one above, such as 3, 3.125 or 3.2. One circle-squarer weighed circles cut out from cardboard to justify his conclusions, claiming that 'algebraical geometry is all moonshine', while another 'misguided visionary'

> filled me with a great ambition — to do a feat I had never yet heard of as accomplished by man, namely to convince a 'Circle-Squarer' of his error! The value my friend had selected for 'π' was *not* an original one — being 3.2: but the enormous error, beginning as early as the *first* decimal place, tempted one with the idea that it could be easily demonstrated to be an error. I should think more than a score of letters were interchanged before I became sadly convinced that I had no chance.

In order to try to convince such enthusiasts of the errors of their ways, Dodgson began to write a book, *Simple Facts about Circle-Squaring*, that would set the record straight. It began as follows:

> Suppose that a controversy had arisen about the details of the battle of Waterloo, and that in a certain Debating Society the question had

been raised as to the exact time when Bulow's Prussian Corps appeared on the field of battle. Disputants, who supported the theory that it was a little before 6 p.m., or a little after, would no doubt be patiently listened to: but what would the Society say to a member who proposed to prove that it took place at 4 p.m. on the *nineteenth* of June? Would they not exclaim with one voice, 'If there is one fact in History more certain than another, it is that the battle was fought on the *eighteenth*. To go outside the limits of that day is simply absurd. We cannot waste our time in listening to any one who does not accept the ordinary *data* of the subject.'

Now this is precisely the position I propose to take with reference to the theories of the "Circle-Squarers," under which term I include all who have attempted to give an *exact* value to the area of a circle, expressed in terms of the square on its radius. The mathematical world are agreed that it is somewhere very near 3.14159 times that square — so near, indeed, that the above number is too small to express it, while 3.1416 is too great. Any one, then, who should suggest the theory that it was a little more or less than this number, say 3.14161 or 3.14158, might perhaps find listeners: but what would be said to a theorist who proposed to prove it to be $4\frac{1}{2}$? "My good sir," we should exclaim, "if there is one fact in Geometry more certain than another, it is that the area of a circle is less than its circumscribed square and greater than its inscribed square; and that these two squares are respectively four times, and twice, the square of the radius. To go outside these limits is simply absurd. We cannot waste our time in listening to any one who does not accept the ordinary *data* of the subject."

Dodgson proposed to settle the matter by presenting simple arguments to convince his readers that

whatever be the exact value of the area, it is at any rate less than 3.1417 times, and greater than 3.1413 times, the square on its radius.

Sadly, the book was never completed, and misguided enthusiasts continue to pester mathematicians around the world with their attempts to square the circle.

Fit the Fifth
Send Me the Next Book . . .

A well-known story relates how Queen Victoria was so utterly charmed by *Alice's Adventures in Wonderland* that she demanded:

> Send me the next book Mr Carroll produces —

The next book duly arrived; it was entitled *An Elementary Treatise on Determinants*. Queen Victoria was not amused.

Dodgson firmly denied this story thirty years later in the second edition of his *Symbolic Logic*:

> I take this opportunity of giving what publicity I can to my contradiction of a silly story, which has been going the rounds of the papers, about my having presented certain books to Her Majesty the Queen. It is so constantly repeated, and is such absolute fiction, that I think it worth while to state, once and for all, that it is utterly false in every particular: nothing even resembling it has ever occurred.

No British newspaper reports have been found that support Dodgson's account, so perhaps it was true after all . . .

In this chapter we describe the writing of *Alice's Adventures in Wonderland* and *An Elementary Treatise on Determinants*, and explain the algebraic ideas behind the latter. But first we return to Charles Dodgson's earlier commitment to train for the priesthood.

Dodgson the Deacon

When Dodgson was awarded a Studentship at Christ Church in 1852, he promised to remain celibate and to proceed to holy orders. The first step was to become a deacon. On 5 August 1861 he wrote to the Diocesan Registrar of Oxford:

> I am intending to offer myself at the Bishop of Oxford's examination in September, to be ordained Deacon.

The Bishop of Oxford was Samuel ('Soapy Sam') Wilberforce, who is best remembered for his part in a celebrated debate with Thomas

AN

ELEMENTARY TREATISE

ON

DETERMINANTS

WITH THEIR APPLICATION TO

SIMULTANEOUS LINEAR EQUATIONS

AND ALGEBRAICAL GEOMETRY.

BY

CHARLES L. DODGSON, M.A.

STUDENT AND MATHEMATICAL LECTURER OF CHRIST CHURCH, OXFORD.

London;

MACMILLAN AND CO.

1867.

Huxley on Darwin's recent theory of evolution during a British Association meeting in Oxford. When the Bishop asked Huxley whether he was descended from an ape on his grandfather's or grandmother's side of the family, Huxley retorted that he would rather have an ape for an ancestor than a bishop.

At first Dodgson had been uncertain whether to take this first step. As he later recalled, in a letter to his cousin and godson,

> When I reached the age for taking Deacon's Orders, I found myself established as the Mathematical Lecturer, and with no sort of inclination to give it up and take parochial work: and I had grave doubts whether it would not be my duty *not* to take orders. I took advice on this point (Bishop Wilberforce was one that I applied to), and came to the conclusion that, so far from educational work (even Mathematics) being unfit occupation for a clergyman, it was distinctly a *good* thing that many of our educators should be men in Holy Orders.

However, he was uncertain whether he should eventually proceed to *priest's* orders:

> I asked Dr. Liddon whether he thought I should be justified in taking Deacon's Orders as a sort of experiment, which would enable me to try how the occupations of clergyman suited me, and *then* decide whether I would take full Orders. He said "most certainly" — and that a Deacon is in a totally different position from a Priest: and much more free to regard himself as *practically* a layman. So I took Deacon's Orders in that spirit.

After weeks of preparation, he duly became a deacon of the Church of England. The ceremony took place on 22 December 1861 in Christ Church Cathedral and was conducted by the Bishop.

In the event, Dodgson never became a priest, having come to believe that it was his duty *not* to do so, since regular parochial duties would take him away from the teaching career to which he felt he had been called. But there were other reasons why he felt unable to proceed.

One was his great love of the theatre, which in his day was widely regarded as a place of ill-repute and no place for a clergyman. When Bishop Wilberforce declared that the 'resolution to attend theatres was an absolute disqualification for Holy Orders' (meaning the priesthood), there was no more to be said.

Another reason was that Dodgson had a speech hesitation. It was less serious than many have claimed, but caused him problems when he was reading in public:

> The hesitation, from which I have suffered all my life, is always worse in *reading* (when I can *see* difficult words before they come) than in speaking. It is now many years since I ventured on reading in public — except now and then reading a lesson in College Chapel. Even that I find such a strain on the nerves that I seldom attempt it. As to reading the *prayers*, there is a much stronger objection than merely my own feelings: every difficulty is an interruption to the devotions of the congregation, by taking off their thoughts from what ought to be the *only* subject in their minds.

Sometimes there were difficulties with certain successions of consonants:

> I got through it all with great success, till I came to read out the first verse of the hymn before the sermon, where the two words 'strife strengthened' coming together, were too much for me, and I had to leave the verse unfinished.

As a result he confined his efforts to preaching sermons, some lasting forty minutes or more, where he could make his own selection of words. According to his nephew, Stuart Dodgson Collingwood,

> his sermons were always delightful to listen to, his extreme earnestness being evident in every word . . . He was essentially a religious man in the best sense of the term, and without any of that morbid sentimentality which is too often associated with the word.

More Pamphlets

Although Dodgson taught mostly geometry, he also taught other areas of pure mathematics. In 1861 he wrote his sixpenny *Notes on the First Part of Algebra*, and followed this later with pamphlets in algebra and trigonometry designed for examination candidates. In *The Formulae of Plane Trigonometry* he proposed new symbols to represent the 'goniometrical ratios' of sine, cosine, tangent, and so on:

sine   cosine   tangent

In February 1861, Dodgson related his current publishing activities in a letter to his sister Mary:

> As you ask me about my mathematical books I will give you a list of my "Works."
> (1) Syllabus, etc. etc. (done)
> (2) Notes on Euclid (done)
> (3) Ditto on Algebra (done — will be out this week, I hope)
> (4) Cycle of examples, Pure Mathematics (about ⅓ done)
> (5) Collection of formulae (½ done)
> (6) Collection of symbols (begun)
> (7) Algebraical Geometry in 4 vols. (about ¼ of Vol. 1 done).
> Doesn't it look grand?

The most unusual of these was his 'Cycle of examples'. In June 1862 Dodgson sent a *Circular to Mathematical Friends* to his Oxford colleagues enclosing 'in compendious form' a list of all the topics in the pure mathematics syllabus for the Finals examination, arranged in twenty-six subject areas and further subdivided into over four hundred topics in total. He also provided a guide to two thousand examples, covering the whole syllabus and organized so that the most important subjects received the greatest number of questions. The result, after many long hours of hard work, was his one-shilling *Guide to the Mathematical Student in Reading, Reviewing, and Working Examples* (1864). The 27-page result was extremely complicated:

> The Cycle intended for this purpose consists of two columns: one containing the numbers from 1 to 1702, the other, references to the Syllabus. It is intended that the student using it should turn to the Syllabus for each reference, and work two or three examples in the subject there indicated . . . and at the end of each day's work mark what point in the Cycle he has reached.

It is doubtful whether anyone ever used it.

The [rebus: nuts]

My [rebus: deer] Ina,

Though [rebus: eye] don't give birthday <u>presents</u>, still [rebus: eye]

April
... write a birthday [rebus: letter].

June
[rebus: ewe] came 2 your [rebus: door] 2 wish U many happy returns of the day, [rebus: butter-barrel] the [rebus: cat] met me, [rebus: hand] took me for a [rebus: rat], [rebus: balloon][rebus: hand] hunted me [rebus: hand] and [rebus: hand] till [rebus: eye] could hardly [rebus: house]

However somehow [rebus: eye] got into the [rebus: house], [rebus: hand] there a [rebus: snake-mouse] met me, [rebus: hand] took me for a [rebus: ball][rebus: mole], and pelted me

Puzzle letter to Georgina Watson

Letters to Child-friends

We have seen that Charles Dodgson was an inveterate letter-writer. Indeed, he wrote and received tens of thousands of letters, both social and professional, and from 1861 until his death in 1898 he kept a register of them all; the last-numbered entry is 98 721.

> I find I write about 20 words a minute, and a page contains about 150 words, i.e. about 7½ minutes to a page. So the copying of 12 pp. took about 1½ hours: and the original writing 2½ hours or more.

Although many letters were to his brothers and sisters or to distinguished personages of the time, the most interesting ones were to children:

> I hardly know which is me and which the inkstand. Pity me, my dear child! The confusion in one's *mind* doesn't so much matter — but when it comes to putting bread-and-butter, and orange marmalade, into the *inkstand*; and then dipping pens into *oneself*, and filling *oneself* up with ink, you know, it's horrid!

Some of his letters were couched in arithmetical language. Here is part of a letter he wrote to his young friend Isa Bowman:

> It's all very well for you & Nellie & Emsie to unite in millions of hugs & kisses, but please consider the *time* it would occupy your poor old very busy Uncle! Try hugging & kissing Emsie for a minute by the watch, & I don't think you'll manage it more than 20 times a minute. "Millions" must mean 2 millions at least.

> | 20 | 2,000,000 | hugs & kisses |
> | --- | --- | --- |
> | 60 | 100,000 | minutes |
> | 12 | 1,666 | hours |
> | 6 | 138 | days [at 12 hours a day |
> | | 23 | weeks. |

I couldn't go on hugging & kissing more than 12 hours a day, & I wouldn't like to spend *Sundays* that way. So you see it would take 23 *weeks* of hard work. Really, my dear Child, I *cannot spare the time* . . .

Please give my kindest regards to your mother, & $\frac{1}{2}$ of a kiss to Nellie, & $\frac{1}{200}$ of a kiss to Emsie, & $\frac{1}{2000000}$ of a kiss to yourself.

Charles Dodgson had a great love of children. His own happy childhood experiences had given him a deep understanding of their minds and an appreciation of their interests. He delighted in writing to them, showing them his puzzles, games and word-play, and generally sharing in their innocent pleasures. One of them recalled

> long walks around Oxford, blissful days in town, and many pleasurable hours spent in the treasure-house of his rooms in Christ Church, where — no matter how often one went — there was always something fresh to be seen, something new and strange to hear.

Although he once jokingly remarked that he was 'fond of children (except boys)', his friendships were with both boys and girls. Here is a letter that he sent to a young lad of fourteen named Wilton Rix, containing an algebraic paradox; the explanation is given in the Notes at the end of the book:

Honoured Sir,

Understanding you to be a distinguished algebraist (i.e. distinguished from other algebraists by different face, different height, etc.), I beg to submit to you a difficulty which distresses me much.

If x and y are each equal to "1," it is plain that $2 \times (x^2 - y^2) = 0$, and also that $5 \times (x - y) = 0$.

Hence $2 \times (x^2 - y^2) = 5 \times (x - y)$.

Now divide each side of this equation by $(x - y)$.

Then $2 \times (x + y) = 5$.

But $(x + y) = (1 + 1)$, i.e. $= 2$.

So that $2 \times 2 = 5$.

Ever since this painful fact has been forced upon me, I have not slept more than 8 hours a night, and have not been able to eat more than 3 meals a day.

I trust you will pity me and will kindly explain the difficulty to

Your obliged, Lewis Carroll.

Sadly, much nonsense has been written about Dodgson's friendships with children. In common with many of his generation, he regarded young children as the embodiment of purity and he delighted in their innocence. His vows of celibacy, which he took extremely seriously, would have outlawed any inappropriate behaviour, and there has never been a shred of evidence of anything untoward. Subjecting him to a modern 'analysis', rather than judging him in the context of his time, is bad history and bad psychology, and often tells us more about the writer than about Dodgson.

Several of Dodgson's earliest attempts at photography took place in the Deanery garden at Christ Church in 1856. It was here that he became acquainted with the Liddell children, Alice, Edith, Lorina and Harry. He took many fine pictures of them, and they quickly became firm friends. Dodgson used to enjoy showing them around Oxford, visiting the colleges and museums and pointing out things of interest.

Boating trips on the river were a particular delight for the children. The most celebrated of these took place on 4 July 1862, when he and his friend the Revd Robinson Duckworth, Fellow of Trinity College, took the Liddell sisters Alice, Edith and Lorina up the river to Godstow; Alice was then aged ten. As Duckworth later recalled:

> I rowed *stroke* and he rowed *bow* in the famous Long Vacation voyage to Godstow, when the three Miss Liddells were our passengers, and the story was actually composed and spoken *over my shoulder* for the benefit of Alice Liddell, who was acting as 'cox' of our gig. I remember turning round and saying, 'Dodgson, is this an extempore romance of yours?' And he replied, 'Yes, I'm inventing as we go along.'
>
> I also well remember how, when we had conducted the three children back to the Deanery, Alice said, as she bade us good-night, 'Oh, Mr Dodgson, I wish you would write out Alice's adventures for me.' He said he should try, and he afterwards told me that he sat up nearly the whole night, committing to a MS. book his recollections of the drolleries with which he had enlivened the afternoon. He added illustrations of his own, and presented the volume, which used often to be seen on the drawing-room table at the Deanery.

The resulting manuscript, entitled *Alice's Adventures Under Ground*, was presented to Alice at Christmas-time 1864. Renamed *Alice's Adventures in Wonderland*, it was published in 1865, and has

Alice, Lorina and Edith Liddell
(photographed by Charles Dodgson, 1860)

Harry Liddell
(photographed by Charles Dodgson, 1859)

The manuscript of Alice's Adventures Under Ground

never been out of print. A facsimile edition of the original manuscript appeared in 1886, and *The Nursery "Alice"*, for younger children, appeared in 1890. The celebrated sequel, *Through the Looking-Glass and What Alice Found There*, was published in 1871.

The *Alice* books contain many allusions to Alice's family and friends; for example:

> the pool was getting quite crowded with the birds and animals that had fallen into it: there was a Duck and a Dodo, a Lory and an Eaglet, and several other curious creatures.

The Duck was Duckworth, the Lory (a type of parrot) was Lorina and the Eaglet was Edith. Dodgson was the Dodo, a stuffed specimen

111

Alice with the Duck, Lory, Eaglet and Dodo

of which he and the Liddell children saw in their regular visits to the new Oxford University Museum.

An Algebra Lesson

In 1867 Dodgson published his most important algebra book, with the snappy title of *An Elementary Treatise on Determinants, with their Application to Simultaneous Linear Equations and Algebraical Geometry.*

The idea of a determinant arises in the context of algebraic geometry (now called 'analytic geometry'), in which algebraic methods are used to obtain results in geometry. This topic was part of the mathematics needed for the Oxford examinations, and in 1860 Dodgson wrote a 164-page book, *A Syllabus of Plane Algebraical*

Geometry, systematically arranged with formal definitions, postulates and axioms. He described this as the 'algebraic analogue' of Euclid's pure geometry; four volumes were projected, but only this first one appeared.

We now explain the mathematical terms that appear in the title of Dodgson's algebra book and introduce his 'method of condensation'. If you are not interested in these, you should skip to the next section, on page 120.

The Algebraic Geometry of the Plane

The use of algebra to solve geometrical problems can be traced back to René Descartes in the seventeenth century. If we draw two axes at right angles in the plane, we can represent each point by a pair (x, y) of numbers, where x and y are the distances of the point from the two axes; the diagram shows the points (4, 1) and (3, 2). Such pairs of numbers (x, y) are now named *Cartesian coordinates*, after Descartes.

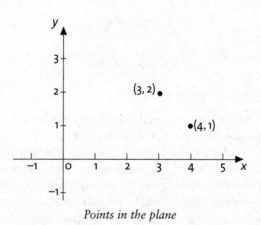

Points in the plane

We can represent lines by *linear equations*. These have the form $ax + by = k$, where a, b and k are numbers; the next diagram shows the lines with equations

$$4x + y = 6 \quad \text{with } a = 4, b = 1, k = 6$$

and

$$3x + 2y = 7 \quad \text{with } a = 3, b = 2, k = 7$$

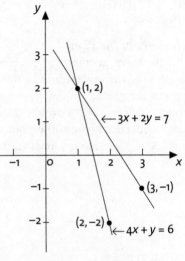

Lines in the plane

The points on the first line are those whose coordinates x and y satisfy the equation $4x + y = 6$; two such points are $(1, 2)$ and $(2, -2)$. The points on the second line are those for which x and y satisfy the equation $3x + 2y = 7$; two such points are $(1, 2)$ and $(3, -1)$.

To find the point where two lines cross, we write down their equations — we call them *simultaneous linear equations* — and use the techniques we learned at school to solve them. In this case, when we solve the simultaneous equations

$$4x + y = 6$$
$$3x + 2y = 7$$

we find that $x = 1$ and $y = 2$. Thus, the two lines cross at the point $(1, 2)$, as shown in the above diagram.

Determinants

Another way of solving simultaneous equations is to use determinants. If a, b, c and d are numbers, their determinant is simply the number

$$(a \times d) - (b \times c) = ad - bc$$

It is written as

$$\begin{vmatrix} a & b \\ c & d \end{vmatrix}$$

and is called a 2 × 2 (read as '2-by-2') *determinant*. Examples are

$$\begin{vmatrix} 4 & 1 \\ 3 & 2 \end{vmatrix} = (4 \times 2) - (1 \times 3) = 8 - 3 = 5$$

$$\begin{vmatrix} 6 & 1 \\ 7 & 2 \end{vmatrix} = (6 \times 2) - (1 \times 7) = 12 - 7 = 5$$

and

$$\begin{vmatrix} 4 & 6 \\ 3 & 7 \end{vmatrix} = (4 \times 7) - (6 \times 3) = 28 - 18 = 10$$

To solve the above simultaneous equations

$$4x + y = 6$$
$$3x + 2y = 7$$

we use these three 2 × 2 determinants, as follows:

- the first determinant involves the four numbers 4, 1, 3, 2 — call it D;
- for the second determinant, we replace the first column $\frac{4}{3}$ of D by the right-hand column $\frac{6}{7}$, and call the result A;
- for the third determinant, we replace the second column $\frac{1}{2}$ of D by the right-hand column $\frac{6}{7}$, and call the result B.

The solution of the equations is then given by $x = A/D$ and $y = B/D$. In this case, using the values $D = 5$, $A = 5$ and $B = 10$ that we calculated above, we have $x = {}^5/_5 = 1$ and $y = {}^{10}/_5 = 2$, giving the point (1, 2), as before.

115

The Algebraic Geometry of Three-dimensional Space

We can extend this process to three dimensions — the details are more complicated, but the basic ideas are the same. We draw three axes at right angles, and represent each point by a triple (x, y, z) of numbers, where x, y and z are the distances of the point from the three axes; the next diagram shows the point $(1, 2, 3)$.

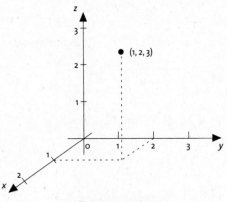

Points in space

We can represent *planes* by linear equations of the form $ax + by + cz = k$, where a, b, c and k are numbers. An example is the plane

$$x + y + z = 4 \quad \text{with } a = b = c = 1, k = 4$$

The points of the plane are those whose coordinates x, y and z satisfy this equation; four such points are $(4, 0, 0)$, $(0, 4, 0)$, $(0, 0, 4)$ and $(2, 1, 1)$.

To find the point where three planes meet, we write down their equations and solve them. Suppose that the three planes are represented by the equations

$$x + 4y + 2z = 8$$
$$x + 2y + 3z = 7$$
$$x + y + z = 4$$

These simultaneous linear equations can be solved to give $x = 2$, $y = 1$, $z = 1$. Thus, the three planes meet at the point $(2, 1, 1)$.

We can also solve these equations by using determinants. We define a 3×3 ('3-by-3') *determinant* in terms of three 2×2 determinants, as follows:

$$\begin{vmatrix} a & b & c \\ d & e & f \\ g & h & i \end{vmatrix} = a\begin{vmatrix} e & f \\ h & i \end{vmatrix} - b\begin{vmatrix} d & f \\ g & i \end{vmatrix} + c\begin{vmatrix} d & e \\ g & h \end{vmatrix}$$

$$= a(ei - fh) - b(di - fg) + c(dh - eg)$$

For example,

$$\begin{vmatrix} 1 & 4 & 2 \\ 1 & 2 & 3 \\ 1 & 1 & 1 \end{vmatrix} = 1 \times \begin{vmatrix} 2 & 3 \\ 1 & 1 \end{vmatrix} - 4 \times \begin{vmatrix} 1 & 3 \\ 1 & 1 \end{vmatrix} + 2 \times \begin{vmatrix} 1 & 2 \\ 1 & 1 \end{vmatrix}$$

$$= (1 \times -1) - (4 \times -2) + (2 \times -1) = -1 + 8 - 2 = 5$$

Similarly,

$$\begin{vmatrix} 8 & 4 & 2 \\ 7 & 2 & 3 \\ 4 & 1 & 1 \end{vmatrix} = (8 \times -1) - (4 \times -5) + (2 \times -1) = 10$$

$$\begin{vmatrix} 1 & 8 & 2 \\ 1 & 7 & 3 \\ 1 & 4 & 1 \end{vmatrix} = (1 \times -5) - (8 \times -2) + (2 \times -3) = 5$$

and

$$\begin{vmatrix} 1 & 4 & 8 \\ 1 & 2 & 7 \\ 1 & 1 & 4 \end{vmatrix} = (1 \times 1) - (4 \times -3) + (8 \times -1) = 5$$

To solve the above simultaneous equations,

$$x + 4y + 2z = 8$$
$$x + 2y + 3z = 7$$
$$x + y + z = 4$$

117

we use these four 3 × 3 determinants, as follows:

- the first determinant involves the nine numbers 1, 4, 2, 1, 2, 3, 1, 1, 1 — call it D;
- for the second determinant, we replace the first column $\begin{smallmatrix}1\\1\\1\end{smallmatrix}$ of D by the right-hand column $\begin{smallmatrix}8\\7\\4\end{smallmatrix}$, and call the result A;
- for the third determinant, we replace the second column $\begin{smallmatrix}4\\2\\1\end{smallmatrix}$ of D by the right-hand column $\begin{smallmatrix}8\\7\\4\end{smallmatrix}$, and call the result B;
- for the fourth determinant, we replace the third column $\begin{smallmatrix}2\\3\\1\end{smallmatrix}$ of D by the right-hand column $\begin{smallmatrix}8\\7\\4\end{smallmatrix}$, and call the result C.

The solution of the equations is then given by $x = A/D$, $y = B/D$, $z = C/D$. Using the values $D = 5$, $A = 10$, $B = 5$ and $C = 5$ that we calculated above, we have $x = {}^{10}/_5 = 2$, $y = {}^5/_5 = 1$ and $z = {}^5/_5 = 1$, giving the point (2, 1, 1), as before.

We can extend these ideas to determinants with more entries, and use them to solve simultaneous linear equations with more than three variables. Unfortunately, the calculation of these larger determinants can be a long and tedious business: for a 4 × 4 determinant we first write it in terms of four 3 × 3 determinants, and then evaluate each of these in terms of three 2 × 2 determinants as before. In 1866 Dodgson broke new ground by discovering an alternative method in which we need to calculate 2 × 2 determinants only.

Dodgson's 'Method of Condensation'
Suppose that we wish to calculate the determinant

$$\begin{vmatrix} 1 & 4 & 2 \\ 1 & 2 & 3 \\ 1 & 1 & 1 \end{vmatrix}$$

(The reason for shading the centre will become clear soon.)

We first calculate the 2 × 2 determinant in each of the four corners:

$$\begin{vmatrix} 1 & 4 \\ 1 & 2 \end{vmatrix} = -2, \quad \begin{vmatrix} 4 & 2 \\ 2 & 3 \end{vmatrix} = 8, \quad \begin{vmatrix} 1 & 2 \\ 1 & 1 \end{vmatrix} = -1, \quad \begin{vmatrix} 2 & 3 \\ 1 & 1 \end{vmatrix} = -1$$

Then we write down the 2 × 2 determinant containing these four numbers:

$$\begin{vmatrix} -2 & 8 \\ -1 & -1 \end{vmatrix} = 10$$

Finally, we divide the result by the shaded number in the middle: $^{10}/_2 = 5$, the correct answer.

So far, nothing has been gained by using Dodgson's method, but it comes into its own when we wish to calculate larger determinants. An example will make the method clear; again, we shade the centre for clarity.

Suppose that we wish to calculate the determinant

$$\begin{vmatrix} 2 & 1 & 1 & 4 \\ 1 & 2 & 1 & 6 \\ 1 & 1 & -2 & -4 \\ 2 & 1 & -3 & -8 \end{vmatrix}$$

We first calculate each of the nine 2 × 2 determinants formed by four neighbouring numbers:

$$\begin{vmatrix} 2 & 1 \\ 1 & 2 \end{vmatrix} = 3, \quad \begin{vmatrix} 1 & 1 \\ 2 & 1 \end{vmatrix} = -1, \quad \begin{vmatrix} 1 & 4 \\ 1 & 6 \end{vmatrix} = 2$$

$$\begin{vmatrix} 1 & 2 \\ 1 & 1 \end{vmatrix} = -1, \quad \begin{vmatrix} 2 & 1 \\ 1 & -2 \end{vmatrix} = -5, \quad \begin{vmatrix} 1 & 6 \\ -2 & -4 \end{vmatrix} = 8$$

$$\begin{vmatrix} 1 & 1 \\ 2 & 1 \end{vmatrix} = -1, \quad \begin{vmatrix} 1 & -2 \\ 1 & -3 \end{vmatrix} = -1, \quad \begin{vmatrix} -2 & -4 \\ -3 & -8 \end{vmatrix} = 4$$

We next write down the 3 × 3 determinant containing these nine numbers, and shade the centre:

$$\begin{vmatrix} 3 & -1 & 2 \\ -1 & -5 & 8 \\ -1 & -1 & 4 \end{vmatrix}$$

We now work with this 3 × 3 determinant. Calculating the 2 × 2 determinant in each corner, we obtain the determinant

$$\begin{vmatrix} -16 & 2 \\ -4 & -12 \end{vmatrix}$$

We next divide each number in this determinant by the corresponding shaded number in the original determinant. This gives

$$\begin{vmatrix} -16/2 & 2/1 \\ -4/1 & -12/-2 \end{vmatrix} = \begin{vmatrix} -8 & 2 \\ -4 & 6 \end{vmatrix} = -40$$

Finally, we divide this result by the shaded number in the centre of the 3 × 3 determinant: −40/−5 = 8, which is the correct answer.

As Dodgson remarked, once you are used to it, you will find that this procedure can be carried out remarkably quickly.

Dodgson's *Determinants*

Determinants have a long history that can be traced back to the seventeenth century. Two hundred years later there were successful textbooks on the subject by the English mathematician and publisher William Spottiswoode and the German mathematician Richard Baltzer, among others. Dodgson was aware of these and referred to both in his book.

Dodgson first mentioned his condensation method in his diary entries for early 1866:

27 *February*: Discovered a process for evaluating arithmetical Determinants, by a sort of condensation, and proved it up to 4^2 terms.

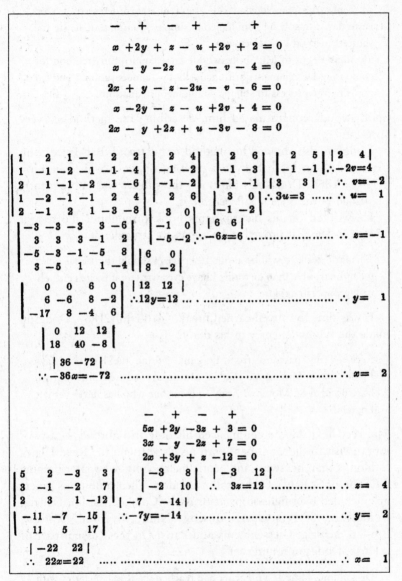

A page from Dodgson's book on determinants

28 February: Completed a rule, built on the process discovered yesterday, for solving simultaneous simple equations. It is far the shortest method I have yet seen.

29 March: Sent to Mr. Spottiswoode an account of my method for computing Determinants arithmetically by "condensation," and for applying the process to solving simultaneous equations.

Spottiswoode congratulated him, describing the method as 'very successful'.

Dodgson also showed his method to his friend 'Bat' Price, who presented it on his behalf at a meeting of the Royal Society on 17 May 1866. Dodgson's paper on the subject was subsequently published in the Society's *Proceedings*.

He worked on his determinants book throughout the rest of 1866 and 1867. It proved to be a massive struggle:

This little book (it will be about 100 pages I should think) has given me more trouble than anything I have ever written: it is such entirely new ground to explore.

It was time for a well-earned break. On 11 July 1867, Dodgson made the following entry in his diary:

Received my passport from London. During the last few days Liddon has informed me that he can go abroad with me, and we have decided on Moscow! Ambitious for one who has never yet left England.

The very next day he commenced his only trip abroad — a two-month visit to the Continent with his close friend the Revd Henry Liddon, a man of 'sweet and gentle melancholy' who later became a Canon of St Paul's Cathedral in London. Their ambitious itinerary included overnight stops at Brussels, Cologne, Berlin, Danzig, Königsberg, St Petersburg, Moscow, Nijni Novgorod, Warsaw, Breslau, Leipzig, Giessen, Ems and Paris. On the Russian part of the trip, Dodgson found that

Dr. Liddon's fame as a preacher had reached the Russian clergy, with the result that he and Mr. Dodgson found many doors open to them which are usually closed to travellers in Russia.

Dodgson's 'Russian diary' is full of extensive descriptions of the churches, cathedrals, palaces and art galleries they visited, from the serious to the inconsequential. While in Danzig he recorded that

> At the hotel was a green parrot on a stand; we addressed it as "Pretty Poll," and it put its head on one side and thought about it, but wouldn't commit itself to any statement. The waiter came up to inform us of the reason of its silence: "Er spricht nicht Englisch; er spricht nicht Deutsch." It appeared that the unfortunate bird could speak nothing but Mexican! Not knowing a word of that language, we could only pity it.

On his return to Oxford, Dodgson completed his book on determinants, which was published in December 1867, on sale at 10s. 6d. In its pages he endeavoured to present the material in a precise and original way, inventing his own notation and terminology when necessary. As with his other mathematical writings, he put his illustrative examples into footnotes so that they did not interrupt the flow of the discussion. His method of condensation appears in an appendix, and the book also includes the first appearance in print of the proof of an important result (now sometimes called the *Kronecker–Capelli theorem*) which had been discovered independently by several people, including himself.

Unfortunately, Dodgson's book was not a success, possibly due to a lack of distribution to key mathematicians, while his terminology and notation were too cumbersome, and his over-formal approach made the book difficult to read.

University Whimsy

As we have seen, there were occasions on which Dodgson's sense of humour extended to University issues. In the early 1860s there was much discussion about the stipend of the Regius Professor of Greek, the Revd Benjamin Jowett. When the Professorship was founded in the sixteenth century the annual salary was £40; three hundred years later it was still £40. Jowett was a highly controversial figure who later became a distinguished Master of Balliol

The Revd Henry Parry Liddon (left) and the Revd Benjamin Jowett

College, but his religious views were regarded by many people (including Dodgson) as heretical. The matter came to a head in 1865, when new salaries of £400 and £500 were proposed.

In March 1865 Dodgson recorded in his diary that

> A day or two ago an idea occurred to me of writing a sham mathematical paper on Jowett's case, taking π to symbolise his payment, and have jotted down a little of it.

The result was *The New Method of Evaluation as Applied to* π, in which Dodgson presented a range of pseudo-mathematical arguments to evaluate π, the salary of J, the Regius Professor of Greek.

The following are the main data of the problem:

Let U = the University, G = Greek, and P = Professor. Then GP = Greek Professor; let this be reduced to its lowest terms, and call the result J.

Also let W = the work done, T = the Times, p = the given payment, π = the payment according to T, and S = the sum required; so that π = S.

The problem is, to obtain a value for π which shall be commensurable with W.

In the early treatises on this subject, the mean value assigned to π will be found to be 40.000000. Later writers suspected that the decimal point had been accidentally shifted, and that the proper value was 400.00000: but, as the details of the process for obtaining it had been lost, no further progress was made in the subject till our own time, though several most ingenious methods were tried for solving the problem.

The pamphlet then presents five 'methods' for evaluating π. One of these was entitled 'Elimination of J':

It had long been perceived that the chief obstacle to the evaluation of π was the presence of J, and in an earlier age of mathematics J would probably have been referred to rectangular axes, and divided into two unequal parts.

It was also proposed to eliminate J by making him a Canon:

The chief objection to this process was, that it would raise J to an inconveniently high power, and would after all only give an irrational value for π.

Eventually, the final value of π = S = 500.00000 was agreed upon:

This result differs considerably from the anticipated value, namely, 400.00000: still there can be no doubt that the process has been correctly performed, and that the learned world may be congratulated on the final settlement of this most difficult problem.

Some years later, in a satirical document on the University's expenditure, Dodgson again referred to Jowett's salary:

Then there's Benjy, again: a nice boy, but I daren't tell you what he costs us in pocket money! Oh, the work we had with that boy till we raised his allowance.

Another example of Dodgson's playful attitude to University affairs was in a letter entitled 'The Offer of the Clarendon Trustees'; the Clarendon Trustees administered the Clarendon Estate, named after the first Earl of Clarendon, a wealthy benefactor to the University. In January 1868 the Professor of Experimental Philosophy, Robert Clifton, sent a letter to the Clarendon Estate proposing the establishment of a new laboratory. Shortly after, Dodgson parodied this letter in putting the case for an appropriately designed Mathematical Institute:

Dear Senior Censor,

In a desultory conversation on a point connected with the dinner at our high table, you incidentally remarked to me that lobster-sauce, "though a necessary adjunct to turbot, was not entirely wholesome."

It is entirely unwholesome. I never ask for it without reluctance: I never take a second spoonful without a feeling of apprehension on the subject of possible nightmare. This naturally brings me on to the subject of Mathematics, and of the accommodation provided by the University for carrying on the calculations necessary in that important branch of Science . . .

It may be sufficient for the present to enumerate the following requisites: others might be added as funds permitted.

A. A very large room for calculating Greatest Common Measure. To this a small one might be attached for Least Common Multiple: this, however, might be dispensed with.

B. A piece of open ground for keeping Roots and practising their extraction: it would be advisable to keep Square Roots by themselves, as their corners are apt to damage others.

C. A room for reducing Fractions to their Lowest Terms. This should be provided with a cellar for keeping the Lowest Terms when found, which might also be available to the general body of Undergraduates, for the purpose of "keeping Terms."

D. A large room, which might be darkened, and fitted up with a magic lantern, for the purpose of exhibiting Circulating Decimals in the act of circulation. This might also contain cupboards, fitted with glass doors, for keeping the various Scales of Notation.

E. A narrow strip of ground, railed off and carefully levelled, for investigating the properties of Asymptotes, and testing practically whether Parallel Lines meet or not: for this purpose it should reach, to use the expressive language of Euclid, "ever so far."

This last process, of "continually producing the Lines," may require centuries or more: but such a period, though long in the life of an individual, is as nothing in the life of the University.

As photography is now very much employed in recording human expressions, and might possibly be adapted to Algebraic Expressions, a small photographic room would be desirable, both for general use and for representing the various phenomena of Gravity, Disturbance of Equilibrium, Resolution, &c., which affect the features during severe mathematical operations.

May I trust that you will give your immediate attention to this most important subject?

Believe me,
Sincerely yours,
MATHEMATICUS.

The Clarendon Laboratory was duly built in 1872, but Oxford University had to wait a further sixty years for a Mathematical Institute.

The Chestnuts, the family home in Guildford

Meat-safes, Majorities and Memory

'. . . and they drew all manner of things — everything that
begins with an M —'
'Why with an M?' said Alice.
'Why not?' said the March Hare.

On 21 June 1868 Charles's father, Archdeacon Dodgson, died suddenly at Croft Rectory. Charles later recalled that

> The greatest blow that has ever fallen on *my* life was the death, nearly thirty years ago, of my own dear father.

As the new head of the large family, Dodgson was now responsible for finding a new home for his many brothers and sisters. After a short search they leased The Chestnuts, a substantial house in Guildford, in Surrey. The family left Croft in early September, and by Christmas were happily settled in their new surroundings.

But family life was not the only change that Dodgson had to deal with at this time. Although his teaching continued much as before, there were other changes in College life and in the direction his studies were taking.

College Life

In October 1868 Dodgson moved into a new suite of rooms. Up to this time, the noblemen who resided in College had first refusal on certain rooms, but a change in the rules resulted in a magnificent suite that had formerly been occupied by the Marquess of Bute becoming available. Dodgson thereby acquired the finest and most expensive accommodation in the College, in the north-west corner of Tom Quad; it included a living room, a dining room and a bedroom. With the College's permission, he constructed on the roof above his rooms a fine photographic studio in which he photographed his child-friends and distinguished personalities of the day. He also purchased some

Dodgson's living room in College

tiles by the artist William De Morgan, son of the Victorian mathematician Augustus De Morgan, for the fireplace in his living room.

Dodgson was now playing an increasingly prominent part in the life of the College. In October 1867, after a great deal of discussion and the passage of The Christ Church Ordinances (Oxford) Bill though Parliament, the College had a new Governing Body — previously all decisions had been made by the Dean and Canons. He thus found himself dealing with such matters as the selection of Senior Students and the appointment of Fellows, and decisions on the College buildings.

> Here's to the Governing Body, whose Art
> (For they're Masters of Arts to a man, Sir!)
> Seeks to beautify Christ Church in every part,
> Though the method seems hardly to answer!

In 1872 it was discovered that the bells in the Cathedral tower were unsafe, and the College proposed to rehouse them in a new belfry over the staircase to the Dining Hall. The new belfry was originally designed as a campanile of wood and copper, but when money ran out and the plan was abandoned, it assumed the shape of a large wooden box which Dodgson likened variously to a tea-caddy, a bathing machine, a clothes horse, a bar of soap and a Greek lexicon. It was eventually clad in stone, to hide the wooden box to which Dodgson took such exception:

> The word "Belfry" is derived from the French *bel*, "beautiful, becoming, meet," and from the German *frei*, "free, unfettered, secure, safe." Thus the word is strictly equivalent to "meat-safe," to which the new belfry bears a resemblance so perfect as almost to amount to coincidence.

'East view of the new Belfry, Ch. Ch. as seen from the Meadow'

Punning on the name of his friend Henry Liddon, Dodgson enquired:

> Was it a Professor who designed this box, which, whether with a lid on or not, equally offends the eye?

and parodied Ariel's song from *The Tempest*:

> Full fathom square the Belfry frowns;
>   All its sides of timber made;
> Painted all in greys and browns;
>   Nothing of it that will fade.
> Christ Church may admire the change —
> Oxford thinks it sad and strange.
> Beauty's dead! Let's ring her knell.
> Hark! now I hear them — ding-dong, bell.

Voting in Elections

In the early 1870s, mainly through College events such as the selection of an architect for the new belfry, Dodgson became involved in elections, ranking procedures and the theory of voting. These topics had been studied extensively by the Marquis de Condorcet around the time of the French Revolution, and Dodgson's original contributions to them, among the most creative of all his mathematical investigations, have been described as the most important after Condorcet's.

There are many systems for voting in an election and of determining the winner once the votes have been cast. In one method, the *simple majority* or first-past-the-post system, the electors have one vote each and cast it for the candidate of their choice; the winner is the candidate with the most votes. In another, the *method of elimination*, each elector lists some or all of the candidates in order of preference; after the votes have been counted, those cast for the least popular candidates are removed or reallocated until a winner emerges.

In all his studies on voting, Dodgson was concerned to achieve complete fairness — both to the winner and also to minority candidates further down the list. In December 1873, he wrote in his diary:

> Began writing a paper (which occurred to me last night) on "Methods of Election," in view of our election of a Lee's Reader in Physics and a Senior Student next Wednesday. The subject ... proved to be much more complicated than I had expected.

In this pamphlet, *A Discussion of the Various Methods of Procedure in Conducting Elections*, he considered several widely used voting systems and constructed ingenious examples to illustrate why each may fail to give the fairest result. Here is his discussion of the two methods mentioned above:

§1. *The Method of a Simple Majority.*

In this Method, each elector names the *one* candidate he prefers, and he who gets the greatest number of votes is taken as the winner. The extraordinary injustice of this Method may easily be demonstrated.

Let us suppose that there are eleven electors, and four candidates, a, b, c, d; and that each elector has arranged in a column the names of the candidates, in the order of his preference; and that the eleven columns stand thus:–

a	a	a	b	b	b	b	c	c	c	d
c	c	c	a	a	a	a	a	a	a	a
d	d	d	c	c	c	c	d	d	d	c
b	b	b	d	d	d	d	b	b	b	b

Here a is considered best by *three* of the electors, and second best by all the rest. It seems clear that he ought to be elected; and yet, by the above method, b would be the winner — a candidate who is considered *worst* by *seven* of the electors! ...

§4. *The Method of Elimination, where the names are voted on all at once.*

In this Method, each elector names the *one* candidate he prefers: the one who gets fewest votes is excluded from further competition, and the process is repeated.

b	b	b	c	c	c	d	d	d	a	a
a	a	a	a	a	a	a	a	a	b	c
d	c	d	b	b	b	c	c	b	d	d
c	d	c	d	d	d	b	b	c	c	b

Here, while b is put last by *three* of the electors, and c and d by *four* each, a is not put lower than second by any. There seems to be no doubt that a's election would be the most generally acceptable: and yet, by the above rule, he would be excluded at once, and ultimately c would be elected.

Dodgson also investigated the *method of marks*, in which each elector is given a specified number of votes that can be assigned in any way to the candidates — all for just one candidate, or divided in any manner among them; the candidate who receives the greatest total of votes is the winner. Dodgson remarked that this method of weighted ranking would be perfect if the electors were all public-spirited enough to select the most generally acceptable candidate, even if that candidate were not one of their own choice, and divided their votes accordingly — but since we are not sufficiently unselfish and would assign all our votes to our favourite candidate, the method is liable in practice to coincide with that of a simple majority which has already been shown to be unsound.

Dodgson concluded with his own modification of the method of marks which he proposed as the best method for future use. In his diary entry for 18 December 1873, he remarks on the result of the College elections:

Election of Baynes and Paget, we partly used my Method.

In the following year Dodgson came to realize that there were difficulties associated with his method of marks, and he was forced to re-examine his approach. His method can be used when one candidate has an absolute majority, but for situations where this is not the case, Dodgson proposed comparing the candidates in pairs, a principle which had been used earlier by Condorcet. It was employed by the Governing Body of Christ Church when choosing which of four architects to entrust with the task of constructing the belfry.

Unfortunately, the process of comparing in pairs can also be inconclusive. For example, suppose that three types of dog food, Wooffo, Doggo and Joocy-Chunks, are tested on three dogs, Rover, Spot and Fido:

- Rover prefers Wooffo to Doggo and Doggo to Joocy-Chunks.
- Spot prefers Doggo to Joocy-Chunks and Joocy-Chunks to Wooffo.
- Fido prefers Joocy-Chunks to Wooffo and Wooffo to Doggo.

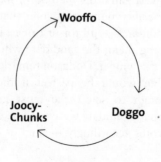

These preferences give rise to a cyclical situation, and in the absence of further information we cannot decide which dog food is the tastiest.

In an election, such cyclical situations become problematic, and in 1876 Dodgson wrote an influential pamphlet, *A Method of Taking Votes on More than Two Issues*, in which he investigated how they can be resolved. He constructed several examples, such as the following one with thirteen electors and four candidates, *a*, *b*, *c* and *d* (ignore the asterisks for the time being):

a	a	a	a	b	b	b	c	c	c	d	d	d
b	b	b	b	d	d	d	d	a	a	b	b	b
c	c	c	c	c	c	c	a*	b	b	c	c	c
d	d	d	d	a	a	a	b*	d	d	a	a	a

In the first-past-the-post system, candidate *a* wins, coming top of the list four times, against only three times for each of the other candidates. But if we now compare candidates *a* and *b*, we find that *a* beats *b* seven times while *b* beats *a* only six times. Similarly, on comparing candidates *b* and *c*, we find that *b* beats *c* ten times while *c* beats *b* only three times. Continuing in this way, we find that, overall, *a* is preferred to *b*, *b* is preferred to *c*, *c* is preferred to *d*, and *d* is preferred to *a* — so we again have a cyclical situation. Who should win the election?

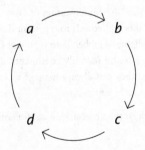

Dodgson drew up a table showing the scores for each pair of candidates, representing the pair *ab* by 7/6 (meaning that *a* beats *b* 7 times while *b* beats *a* 6 times) and the pair *bc* by 10/3. He obtained the following list:

$$ab = 7/6, \quad ac = 4/9, \quad ad = 6/7, \quad bc = 10/3, \quad bd = 9/4, \quad cd = 7/6$$

He then observed that candidate b is preferred to both c and d, and would also be preferred to a if the eighth elector agreed to interchange the asterisked preferences a and b; thus, with *only one* change of vote, ab becomes $6/7$ and b becomes the outright winner. However, a would require *four* changes of vote to become the winner — three against c (to change $4/9$ to $7/6$) and one against d (to change $6/7$ to $7/6$) — and c and d would similarly need *four* changes of vote. Dodgson thereby claimed that 'candidate b is most worthy of the prize, rather than candidate a'.

Memoria Technica

In the mid-1870s, Charles Dodgson resolved to develop his already remarkable powers of memory. To this end he devised a *Memoria Technica* with whose help he and his friends could recall dates and other numbers. His method, dating from 1875, improves on one introduced in 1730 by Dr Richard Grey, a Northamptonshire rector, and is based on the idea of replacing each number by a succession of letters, and then combining these letters into words expressed in the form of a verse that can easily be remembered.

Memoria Technica.

My "Memoria Technica" is a modification of Gray's [*sic*]: but, whereas he used both consonants and vowels to represent digits, and had to content himself with a syllable of gibberish to represent the date or whatever other number was required, I use only consonants, and fill in with vowels "ad libitum", and thus can always manage to make a real word of whatever number has to be represented.

The principles, on which the necessary 20 consonants have been chosen, are as follows:–

    [1] "*b*" and "*c*", the first two consonants in the Alphabet.
    [2] "*d*" from "duo"; "*w*" from "two".
    [3] "*t*" from "tres"; the other may wait awhile.
    [4] "*f*" from "four"; "*q*" from "quatuor".

---

[5] "*l*" and "*v*", because "L" and "V" are the Roman symbols for "fifty" and "five".

[6] "*s*" and "*x*", from "six".

[7] "*p*" and "*m*", from "septem".

[8] "*h*" from "huit"; & "*k*" from the Greek "okto".

[9] "*n*" from "nine"; & "*g*", because it is so like a "9".

[10] "*z*" and "*r*", from "zero".

There is now one consonant still waiting for its digit, viz. "*j*"; and one digit waiting for its consonant, viz. "3": the conclusion is obvious.

The result may now be tabulated thus

1	2	3	4	5	6	7	8	9	0
b	*d*	*t*	*f*	*l*	*s*	*p*	*h*	*n*	*z*
c	*w*	*j*	*q*	*v*	*x*	*m*	*k*	*g*	*r*

When a word has been found, whose last consonants represent the number required, the best plan is to put it as the last word of a rhymed couplet, so that, whatever other words in it are forgotten, the rhyme will secure the only really important word.

Now suppose that you wish to remember the date of the discovery of America, which is "1492": the "1" may be left out as obvious: all we need is "492".

Write it thus:–

4	9	2
f	*n*	*d*
q	*g*	*w*

and try to find a word that contains "*f*" or "*q*", "*n*" or "*g*", "*d*" or "*w*". A word soon suggests itself — "*found*".

The poetical faculty must now be brought into play, and the following couplet will soon be evolved:–

"Columbus sailed the world around,
Until America was FOUND".

If possible, invent the couplet for yourself: you will remember them better than any made by others.

Using his table, Dodgson constructed an ingenious couplet that gives the years of accession of the eight King Henrys of England:

Crazy belief we cause to none
A fact if dead of a hilly run.

Here the consonants are

c r z / b l f / w c s / t n n / f c t / f d d / f h l / l r n

Translating these into numbers and supplying the initial 1 in each case gives the dates 1100, 1154, 1216, 1399, 1413, 1422, 1485 and 1509.

Similar examples were recalled by Evelyn Hatch, one of his child-friends:

I possess some notes in his own handwriting giving rhymes for the dates of the Oxford colleges, as follows:–

Ch. Ch. *Ring Tom when you please*
We ask but small fees.

B. N. C. *With a nose that is brazen*
Our gate we emblazon.

Each of the last three consonants denotes a number, *l* for 5, *f* for 4, and *s* for 6, so that the date for Christ Church is 1546, while in the case of Brazenose *l*, *z*, *n*, give 1509, it being always taken for granted that 1 is the first figure.

Dodgson extended these ideas to various mathematical numbers. In October 1875 he recorded in his diary that he

Sat up until nearly 2 making a "Memoria Technica" for Baynes for logarithms of primes up to 41. I can now calculate in a few minutes almost any logarithm without book.

His verse for log 2 = 0.3010300 was:

Two jockeys to carry
Made *that* racer tarry.

The final seven consonants, *t r c r t r r*, give the numbers 3, 0, 1, 0, 3, 0, 0.

Dodgson then extended his list to the logarithms of the primes up to 100 and beyond, and developed verses to help him remember the first 71 digits of π. He considered writing a book on fast mental calculation, entitled *Logarithms by Lightning: A Mathematical Curiosity*, but it never materialized. He extended his techniques to trigonometry, calculating the values of sin 18° (~0.309) and sin 73° 20⫢ (~0.95) in his head, and recorded that it took him just nine minutes to work out the 13th root of 87 654 327 and fourteen minutes to calculate π^π.

Endings and Beginnings

In 1880 the College was suffering a financial crisis. At the beginning of February, Dodgson wrote to the College authorities with a generous offer for saving money:

> The idea occurred to me that it would be right to lay before the "Committee of Salaries" the history of the work and pay of the Lectureship. The work is now so light that I think the pay may fairly be reduced.

He duly wrote to the Dean proposing that his salary should be lowered from £300 per year to £200. A month later he remarked that

> My work has been absurdly light this term, fully confirming me as to my having done the right thing in offering to resign to the House £100 a year of my present stipend.

In 1881 Dodgson decided to give up the Mathematical Lectureship he had held for twenty-five years to devote more time to his books and articles. In July he wrote:

> My chief motive for holding on has been to provide money for others (for myself, I have been for many years able to retire) but even the £300 a year I shall thus lose I may fairly hope to make by the additional time I shall have for book-writing. I think of asking the Governing Body, next term, to appoint my successor so that I may retire at the end of the year, when I shall be close on 50 years old, and shall have held the Lectureship for exactly 26 years.

On 18 October he resigned in order to do more writing,

> partly in the cause of Mathematical education, partly in the cause of innocent recreations for children, and partly, I hope (though so utterly unworthy of being allowed to take up such work) in the cause of religious thought.

Three days later he received a reply from the Dean promising that

> arrangements should be made, as far as could be done, to carry out my wishes: and kindly adding an expression of regret at losing my services, but allowing that I had "earned a right to retirement." So my Lectureship seems to be near its end.

Dodgson's final official lecture was at the end of November:

> This morning I have given what is most probably my *last*: the lecture is now reduced to nine, of whom all attended on Monday: this morning being a Saint's Day, the attendance was voluntary, and only two appeared, E. H. Morris and G. Lavie. I was lecturer when the *father* of the latter took his degree, *viz.* in 1858.

Almost exactly one year later, the Curator of the Christ Church Common Room resigned, having held the position for twenty-one years. Dodgson reluctantly agreed to take on the job:

> I was proposed by Holland, and seconded by Harcourt, and accepted office with no light heart: there will be much trouble and thought needed to work it satisfactorily: but it will take me out of myself a little, and so may be a real good. My life was tending to become too much that of a selfish recluse.

The job was a time-consuming one, involving the day-to-day management of the Common Room. Dodgson would hold the office of Curator for nine years.

His decision to give up the Lectureship in order to devote more time to his writings seems to have been justified. A diary entry for 1885 records:

> Never before have I had so many literary projects on hand at once. For curiosity I will here make a list of them:
>
> (1) *Supplement to Euclid and His Modern Rivals*, now being set up in pages. This will contain the review of Henrici, and

extracts from reviews of *E.&M.R.* with my remarks on them. I think of printing 250.

(2) Second edition of *Euclid and His Modern Rivals*, this I am correcting for press, and shall embody above in it.

(3) A book of Mathematical curiosities, which I think of calling *Pillow Problems, and other Mathematical Trifles.* This will contain Problems worked out in the dark, Logarithms without Tables, Sines and Angles ditto, a paper which I am now writing, on "Infinities and Infinitesimals," condensed Long Multiplication, and perhaps others.

(4) *Euclid V*, treating Incommensurables by a method of Limits, which I have nearly completed.

(5) *Plain Facts for Circle-Squarers* which is nearly complete, and gives actual proof of limits 3.14158, 3.14160.

(6) A symbolical Logic, treated by my algebraic method.

(7) *A Tangled Tale*, with answers, and perhaps illustrated by Mr. Frost.

(8) A collection of Games and Puzzles of my devising, with fairy-pictures by Miss E. G. Thomson. This might also contain my "Memoria Technica" for dates etc., my "Cipher-writing," scheme for Letter-registration, etc. etc.

(9) Nursery "Alice," for which 20 pictures are now being coloured by Mr. Tenniel.

(10) Serious poems in Phantasmagoria. I think of calling it "Reason and Rhyme," and hope to get Mr. Furniss to draw for it.

(11) *Alice's Adventures Under Ground*, a facsimile of the MS. book, lent me by "Alice" (Mrs. Hargreaves). I am now in correspondence with Dalziel about it.

(12) *Girls' Own Shakespeare*. I have begun on *Tempest*.

(13) New edition of *Parliamentary Representation*, embodying supplement etc.

(14) New edition of *Euclid I, II*, for which I am now correcting edition 4.

(15) The new child's book, which Mr. Furniss is to illustrate: he now has "Peter and Paul" to begin on. I have settled on no name as yet, but it will perhaps be *Sylvie and Bruno*.

I have other shadowy ideas, e.g. a Geometry for Boys, a volume of Essays on theological points freely and plainly treated, and a drama on *Alice* (for which Mr. Mackenzie would write music): but the above is a fair example of "too many irons in the fire"!

Lawn Tennis Tournaments

In 1883 Dodgson became interested in the mathematics of singles tennis tournaments. In a *round-robin tournament*, each of the players plays against all the others, and their scores are tallied in order to find the winner. In a *knockout tournament*, the players compete once in each round, with half the players eliminated in each round, until the winner eventually emerges. But in the latter competition format, as Dodgson remarked, the next-best players in the tournament may not always finish near the top of the list:

> At a Lawn Tennis Tournament, where I chanced, some while ago, to be a spectator, the present method of assigning prizes was brought to my notice by the lamentations of one of the Players, who had been beaten (and had thus lost all chance of a prize) early in the contest, and who had had the mortification of seeing the 2nd prize carried off by a Player whom he knew to be quite inferior to himself.

A game of tennis

To illustrate this, imagine a knockout tennis tournament with eight players, ranked in order of merit with player 1 as the best and player 8 as the worst, and that each game goes according to form. Let us take the worst-case scenario in which in the first round 1 plays 2, 3 plays 4, 5 plays 6 and 7 plays 8. The winners are then 1, 3, 5 and 7, those winning the second round are 1 and 5, and the first prize is then won, appropriately, by player 1. However, the second

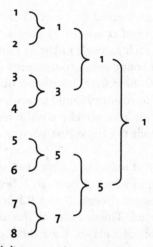

Scheduling a tennis tournament

prize, which should have been awarded to player 2, is awarded instead to player 5, who started in the lower half of the ranking.

Dodgson carried out a similar analysis for a knockout tournament with 32 players, ranked in order of merit. In this case the second prize would be awarded to player 17 instead of player 2, and the third and fourth prizes would be awarded to players 9 and 25 instead of players 3 and 4.

In his pamphlet *Lawn Tennis Tournaments: The True Method of Assigning Prizes with a Proof of the Fallacy of the Present Method*, written under his pen-name of Lewis Carroll, he attempted to circumvent this difficulty:

> The results of the investigations, which I was led to make, I propose to lay before the reader under the following four headings:–

(a) A proof that the present method of assigning prizes is, except in the case of the first prize, entirely unmeaning.

(b) A proof that the present method of scoring in matches is constantly liable to lead to unjust results.

(c) A system of rules for conducting Tournaments, which, while requiring even less time than the present system, shall secure equitable results.

(d) An equitable system for scoring in matches.

Dodgson first observed that every knockout tennis tournament with 32 players must consist of 31 games, since exactly one player is eliminated in each game. It might be fairer to decide the winners by organizing a round-robin tournament in which each of the 32 players plays *all the others*, but that would require 496 separate games, which is unrealistic. Could he achieve a compromise? Could he devise a tennis tournament with a reasonably small number of games in which the three best players emerge in the top three positions?

A simple way of reducing the number of matches is to eliminate 'cyclical triples', in which A beats B, B beats C, and C beats A; in Dodgson's system, if A beats B and B beats C, then the remaining game is not played. This led him to the idea of a player's 'superiors': the superiors of a player X are those players who have beaten X or who have beaten someone else who has beaten X.

Based on this idea, Dodgson first arranged the 32 players in 16 pairs. After the first round he drew up a list of each player's superiors. For the second round he arranged the 16 winners of the first round in pairs, and also the 16 losers, and in later rounds he organized the pairing similarly, first pairing the unbeaten, then those with just one superior, and so on. At each stage any player with three superiors was removed from the list.

Dodgson's method had the desired effect. It required the playing of only 61 games, and ensured that the top three prizes go to the best three players. Although his method was not foolproof, it was a great advance on anything else available at the time. In fact, in his analysis of cyclical situations, Dodgson was way ahead of his time; the subject of ranking in tournaments was not seriously taken up again until the 1940s.

Parliamentary Representation

The period after the 1880 General Election was an interesting time in British politics. There were two main parties — the Liberals, who had been returned to power with William Ewart Gladstone as Prime Minister, and the Conservatives, led by Lord Salisbury after the death of Benjamin Disraeli in 1881.

Much had changed over the previous fifty years. The Reform Acts of 1832 and 1867 had extended the franchise, first to merchants and then to industrial workers, while the Ballot Act of 1872 had allowed secret voting for the first time. The new Liberal government extended the franchise still further, from about 2.6 million electors to over 4.3 million, by giving the vote to agricultural workers and miners, a measure contained in the Third Reform Bill of 1884.

Of the 658 Members of Parliament elected in 1880, fewer than 200 represented single-member electoral districts. More than 400 were in two-member constituencies, while a handful of electoral districts returned three or four members. The Third Reform Bill proposed a massive increase in the number of single-member districts, and at the 1885 General Election no fewer than 616 of the 670 members became the sole representatives of their constituents.

In his desire to see both majority and minority views represented in Parliament, Dodgson was bitterly opposed to these single-member districts. In electoral districts returning members from different parties, both majority and minority views could be fairly represented, but in a single-member district with roughly equal numbers of supporters from each side, the views of almost half the electorate had no voice in Parliament. He was also concerned that if there were too many single-member districts, then a disproportionate number of seats would go to the larger Liberal Party at the expense of the less popular Conservative Party, which he supported. As he noted in 1881,

> So long as general elections are conducted as at present we shall be liable to oscillations of political power, like those of 1874 and 1880, but of ever-increasing violence — one Parliament wholly at the mercy of one political party, the next wholly at the mercy of the other — while the Government of the hour, joyfully hastening to

undo all that its predecessors have done, will yield a majority so immense that the fate of every question will be foredoomed, and debate will be a farce . . .

This concern has been borne out in recent years, with political parties gaining large majorities on a relatively small proportion of the total vote.

Dodgson's own preference was for electoral districts with two to five members, in which electors were given just one vote. For such multi-member constituencies Dodgson calculated the percentages of the votes required for a political party to return a specified number of members, and presented his results in the following table:

| | | number of seats to fill | | | | | |
		1	2	3	4	5	6
	1	**51**					
number	2	34	**67**				
of members	3	26	51	76			
allocated to	4	21	41	61	**81**		
district	5	17	34	51	67	84	
	6	15	29	43	58	72	86

For example, in a single-member district a party requires over half the votes in order to fill the seat — that is, at least 51 per cent (line 1, column 1), if we stick to whole numbers. In order to return two members, a party requires at least 67 per cent of the vote (line 2, column 2) in a two-member district, and 41 per cent (line 4, column 2) in a four-member one. In general, as Dodgson discovered, a political party wishing to return k members in an n-member district requires more than $k/(n + 1)$ per cent of the vote; for example, in order to fill two seats ($k = 2$) in a four-member district ($n = 4$), a party needs more than two-fifths of the vote — that is, at least 41 per cent. The last number in each line gives the percentage of the electorate represented by their members; for example, in a four-member district, 81 per cent of the electorate is so represented.

In 1884 the Proportional Representation Society, seeking the best way of allocating seats in multi-member constituencies, proposed a reform based on the single transferable vote. In this system the voters indicate their first, second and subsequent choices of candidate. As soon as a candidate amasses enough votes, he is declared elected and his surplus votes are then redistributed among the other candidates according to the electors' other preferences.

Dodgson had other ideas. He wrote a number of letters proposing various methods of Parliamentary reform to Lord Salisbury, a fellow Christ Church undergraduate from the early 1850s, and to all the other Members of Parliament. In November 1884 he published *The Principles of Parliamentary Representation*, his major work on the subject, in which he coordinated several ideas that he had been contributing to the *St. James Gazette*. This pamphlet was later described enthusiastically by the political economist Duncan Black as 'the most interesting contribution to Political Science that has ever been made'.

Dodgson's pamphlet began with a list of desiderata, which can be summarized as follows:

- Each elector should have the same chance of being represented in the House of Commons.
- Each Member of Parliament should represent the same number of voters, and this should be uniform throughout the Kingdom.
- The number of unrepresented electors should be as small as possible.
- The proportions of political parties in the House of Commons should be, as nearly as possible, the same as in the country as a whole.
- The process of voting, and of counting the votes and announcing the result, should be as simple as possible.
- The waste of votes caused by more votes being given for a candidate than are needed for his return, should be as far as possible prevented.
- The electors in each district should be, as far as possible, uninfluenced by the results of elections in other districts.

After a great deal of detailed calculation and analysis, Dodgson came to the following conclusions:

- That electoral Districts should be so large as to return, on an average, 3 or more Members each: and that single-Member Districts should be, as far as possible, done away with.
- That Members should be assigned to the several Districts in such numbers that the quota, needed to return a Member, should be tolerably uniform throughout the Kingdom.
- That each Elector should give one vote only.
- That all votes given should be at the absolute disposal of the Candidate for whom they are given, whether to use for himself, or to transfer to other Candidates, or to leave unused.
- That the Elections in the several Districts should terminate, as nearly as possible, at the same time.

He also recommended that there should be 660 members divided among 180 electoral districts, and that no district should contain fewer than 60000 or more than 500000 voters. He drew up a table showing how each electoral district should be allocated a fixed number of members according to its size — for example, while a single-member district would have 60 000 to 105 000 electors, an electoral district with 195 000 to 240 000 voters should be represented by four members.

Some of Dodgson's political recommendations were eventually adopted in England, such as the rule that no results can be announced until all polling stations have closed. Others, such as his various proposals for proportional representation, were not. In the 1870s Dodgson had made known his intention to write a book on voting and elections. Such a publication never materialized, causing the Oxford philosopher Michael Dummett to remark, many years later:

> It is a matter of the deepest regret that Dodgson never completed the book that he planned to write on the subject. Such were the lucidity of exposition and his mastery of the topic that it seems possible that, had he ever published it, the political theory of Britain would have been significantly different.

An intriguing thought indeed!

Fit the Seventh
Puzzles, Problems and Paradoxes

After *Through the Looking-Glass*, Charles Dodgson continued to produce books for children, using his pen-name of Lewis Carroll. In 1876 he wrote his extended poem *The Hunting of the Snark*, which originated in the following manner:

> I was walking on a hillside, alone, one bright summer day, when suddenly there came into my head one line of verse — one solitary line — "For the Snark *was* a Boojum, you see." I knew not what it meant, then: I know not what it means, now; but I wrote it down: and some time afterwards, the rest of the stanza occurred to me, that being its last line: and so by degrees, at odd moments during the next year or two, the rest of the poem pieced itself together, that being its last stanza.

> In the midst of the word he was trying to say,
>   In the midst of his laughter and glee,
> He had softly and suddenly vanished away —
>   For the Snark *was* a Boojum, you see.

Shortly after the poem appeared, Carroll quizzed one of his child-friends, Birdie (Florence Balfour):

> When you have read the *Snark*, I hope you will write me a little note and tell me how you like it, and if you can *quite* understand it. Some children are puzzled with it. Of course you know what a Snark is? If you do, please tell *me*: for I haven't an idea what it is like.

As well as writing books for his friends, Carroll entertained them with his latest puzzles and paradoxes. In this chapter we look at some of these, beginning with a book of mathematical posers.

'Pursuing the Snark with forks and hope'

A Tangled Tale

From April 1880 to November 1884 Lewis Carroll wrote a puzzle column for a periodical called *The Monthly Packet*. Each issue featured a story that concealed some ingenious mathematical problems. The ten stories, called 'Knots', were subsequently collected together into a puzzle book, *A Tangled Tale*, which appeared in time for Christmas 1885 and cost 4s. 6d.

> "A knot!" said Alice, always ready to make herself useful, and looking anxiously about her. "Oh, do let me help to undo it!"

The book was dedicated 'To My Pupil':

> Beloved Pupil! Tamed by thee,
> Addish-, Subtrac-, Multiplica-tion,
> Division, Fractions, Rule of Three,
> Attest thy deft manipulation!
>
> Then onward! Let the voice of Fame
> From Age to Age repeat thy story,
> Till thou hast won thyself a name
> Exceeding even Euclid's glory!

The pupil's name is spelt out by the second letter of each line of this poem: Edith Rix was the sister of Wilton Rix, the fourteen-year-old to whom he wrote his algebra letter (see Fit the Fifth). Miss Rix went on to study mathematics at Cambridge, even though Dodgson had tried to persuade her to do so at Oxford.

The Knots to be undone had such unlikely names as 'Excelsior', 'Her Radiancy', 'A Serpent with Corners' and 'Chelsea Buns'. The author explained that his purpose was

> to embody in each Knot (like the medicine so dexterously, but ineffectually, concealed in the jam of our early childhood) one or more mathematical questions — in Arithmetic, Algebra, or Geometry, as the case might be — for the amusement, and possible edification, of the fair readers of that Magazine.

Knot 1 is the shortest of the ten tangled tales: we present it here in full.

Knot 1: Excelsior

'Goblin, lead them up and down.'

The ruddy glow of sunset was already fading into the sombre shadows of night, when two travellers might have been observed swiftly — at a pace of six miles in the hour — descending the rugged side of a mountain; the younger bounding from crag to crag with the agility of a fawn, while his companion, whose aged limbs seemed ill at ease in the heavy chain armour habitually worn by tourists in that district, toiled on painfully at his side.

As is always the case under such circumstances, the younger knight was the first to break the silence.

'A goodly pace, I trow!' he exclaimed. 'We sped not thus in the ascent!'

'Goodly, indeed!' the other echoed with a groan. 'We clomb it but at three miles in the hour.'

'And on the dead level our pace is —' the younger suggested; for he was weak in statistics, and left all such details to his aged friend.

'Four miles in the hour,' the other wearily replied. 'Not an ounce more,' he added, with that love of metaphor so common in old age, 'and not a farthing less!'

''Twas three hours past high noon when we left our hostelry,' the young man said, musingly. 'We shall scarce be back by supper-time. Perchance mine host will roundly deny us all food!'

'He will chide our tardy return,' was the grave reply, 'and such a rebuke will be meet.'

'A brave conceit!' cried the other, with a merry laugh. 'And should we bid him bring us yet another course, I trow his answer will be tart!'

'We shall but get our deserts,' sighed the elder knight, who had never seen a joke in his life, and was somewhat displeased at his companion's untimely levity. ''Twill be nine of the clock,' he added in an under tone, 'by the time we regain our hostelry. Full many a mile shall we have plodded this day!'

'How many? How many?' cried the eager youth, ever athirst for knowledge.

The old man was silent.

'Tell me,' he answered, after a moment's thought, 'what time it was when we stood together on yonder peak. Not exact to the minute!' he added hastily, reading a protest in the young man's face. 'An' thy guess be within one poor half-hour of the mark, 'tis all I ask of thy mother's son!

Then will I tell thee, true to the last inch, how far we shall have trudged betwixt three and nine of the clock.'

A groan was the young man's only reply; while his convulsed features, and the deep wrinkles that chased each other across his manly brow, revealed the abyss of arithmetical agony into which one chance question had plunged him.

'At a pace of six miles in the hour'

In the *Monthly Packet*, each Knot appeared in story form in one issue, and the problems were summarized, answered and discussed in the following one. For *Knot I*, Carroll recast the question as follows:

> *Problem.* Two travellers spend from 3 o'clock till 9 in walking along a level road, up a hill, and home again: their pace on the level being 4 miles an hour, up hill 3, and down hill 6. Find distance walked: also (within half an hour) time of reaching top of hill.
>
> *Answer.* 24 miles: half past 6.
>
> *Solution.* A level mile takes $1/4$ of an hour, up hill $1/3$, down hill $1/6$. Hence to go and return over the same mile, whether on the level or on the hillside, takes $1/2$ an hour. Hence in 6 hours they went 12 miles out and 12 back. If the 12 miles out had been nearly all level, they would have taken a little over 3 hours; if nearly all up hill, a little under 4. Hence $3^{1}/_{2}$ hours must be within $1/2$ an hour of the time taken in reaching the peak; thus, as they started at 3, they got there within $1/2$ an hour of $1/2$ past 6.

He then presented a summary of the answers he had received:

> Twenty-seven answers have come in. Of these, 9 are right, 16 partially right, and 2 wrong. The 16 give the *distance* correctly, but they have failed to grasp the fact that the top of the hill might have been reached at *any* moment between 6 o'clock and 7.

The solutions had been submitted under pseudonyms, such as FIFEE, A MOTHER'S SON, A SOCIALIST, VIS INERTIAE and YAK. After analysing each solution in turn, Carroll ranked the successful entrants in a 'Class list':

CLASS LIST

I

A MARLBOROUGH BOY.	PUTNEY WALKER.

II

BLITHE.	ROSE.
E. W.	SEA-BREEZE.
L. B.	SIMPLE SUSAN.
O. V. L.	MONEY-SPINNER.

One correspondent, SCRUTATOR, later wrote in to complain that the second question in Knot I 'answers itself'. Carroll responded:

> It is interesting to know that the question "answers itself," and I am sure it does the question great credit: still I fear I cannot enter it on the list of winners, as this competition is only open to human beings.

The remaining Knots are listed below, with Carroll's summaries of the mathematical problems hidden in them; his answers to these problems are given in the Notes at the end of the book.

Knot II, 'Eligible Apartments', features two problems, concealed as before in a story: one problem is about a dinner party and the other concerns the numbering of houses around a square.

> *Problem 1.* The Governor of Kgovjni wants to give a very small dinner party, and invites his father's brother-in-law, his brother's father-in-law, his father-in-law's brother, and his brother-in-law's father. Find the number of guests.
> *Problem 2.* A Square has 20 doors on each side, which contains 21 equal parts. They are numbered all round, beginning at one corner. From which of the four, Nos. 9, 25, 52, 73, is the sum of the distances, to the other three, least?

Knot III, 'Mad Mathesis', concerns two travellers on a circular railway.

> *Problem*: (1) Two travellers, starting at the same time, went opposite ways round a circular railway. Trains start each way every 20 minutes, the easterly ones going round in 3 hours, the westerly in 2. How many trains did each meet on the way, not counting trains met at the terminus itself.
> (2) They went round, as before, each traveller counting as 'one' the train containing the other traveller. How many did each meet?

Knot IV, 'The Dead Reckoning', involves the weighing of sacks during a sea voyage.

> *Problem.* There are 5 sacks, of which Nos. 1, 2, weigh 12 lbs.; Nos. 2, 3, $13\frac{1}{2}$ lbs.; Nos. 3, 4, $11\frac{1}{2}$ lbs.; Nos. 4, 5, 8 lbs.; Nos. 1, 3, 5, 16 lbs. Required the weight of each sack.

Knot V, 'Oughts and Crosses', concerns the assignment of up to three marks to each of ten pictures exhibited in the Royal Academy.

Problem: To mark pictures [with or ○], giving 3 's to 2 or 3, 2 to 4 or 5, and 1 to 9 or 10; also giving 3 ○'s to 1 or 2, 2 to 3 or 4 and 1 to 8 or 9; so as to mark the smallest possible number of pictures, and to give them the largest possible number of marks.

Knot VI, 'Her Radiancy', features two problems, one about money and the other on comparing scarves.

Problem 1: A and B begin the year with only £1,000 a-piece. They borrowed nought; they stole nought. On the next New-Year's Day they had £60,000 between them. How did they do it?
Problem 2: L makes 5 scarves, while M makes 2: Z makes 7 while L makes 3. Five scarves of Z's weigh one of L's; 5 of M's weigh 3 of Z's. One of M's is as warm as 4 of Z's: and one of L's as warm as 3 of M's. Which is best, giving equal weight in the result to rapidity of work, lightness, and warmth?

Knot VII, 'Petty Cash', presents an algebra problem about lemonade, sandwiches and biscuits.

Problem: Given that one glass of lemonade, 3 sandwiches, and 7 biscuits, cost 1s. 2d.; and that one glass of lemonade, 4 sandwiches, and 10 biscuits, cost 1s. 5d.: find the cost of (1) a glass of lemonade, a sandwich, and a biscuit; and (2) 2 glasses of lemonade, 3 sandwiches, and 5 biscuits.

Knot VIII, 'De Omnibus Rebus', presents two problems: one is about pigs in sties and the other, entitled 'The Grurmstipths', is on the timing of omnibuses.

Problem 1: Place twenty-four pigs in four sties so that, as you go round and round, you may always find the number in any sty nearer to ten than the number in the last.
Problem 2: Omnibuses start from a certain point, both ways, every 15 minutes. A traveller, starting on foot along with one of them, meets one in 12½ minutes: when will he be overtaken by one?

Knot IX, 'A Serpent with Corners', poses three problems, two on the displacement of water and the other about an oblong garden.

> *Problem 1*: Lardner states that a solid, immersed in a fluid, displaces an amount equal to itself in bulk. How can this be true of a small bucket floating in a larger one?
>
> *Problem 2*: Balbus states that if a certain solid be immersed in a certain vessel of water, the water will rise through a series of distances, two inches, one inch, half an inch, &c., which series has no end. He concludes that the water will rise without limit. Is this true?
>
> *Problem 3*: An oblong garden, half a yard longer than wide, consists entirely of a gravel-walk, spirally arranged, a yard wide and 3,630 yards long. Find the dimensions of the garden.

Finally, Knot X, 'Chelsea Buns', features problems about the injuries sustained by Chelsea pensioners and about the ages of three sons; another one, similar to 'Where does the day begin?' (see Fit the First), was not answered by the author.

> *Problem 1*: If 70 per cent. of Chelsea Pensioners have lost an eye, 75 per cent. an ear, 80 per cent. an arm, 85 per cent. a leg: what percentage, at least, must have lost all four?
>
> *Problem 2*: At first, two of the ages are together equal to the third. A few years afterwards, two of them are together double of the third. When the number of years since the first occasion is two-thirds of the sum of the ages on that occasion, one age is 21. What are the other two?

Carroll's Puzzles

We now present a few of the many puzzles and problems with which Carroll entertained his friends, young and old. The answers to selected ones appear in the Notes at the end of the book.

Arithmetical Puzzles

An arithmetical curiosity which was popular in Victorian times, and which Carroll enjoyed showing to his child-friends, featured the 'magic number' 142857, which he asked them to multiply by 2, 3, 4, 5, 6 and 7:

A Magic Number: 142857

285714	twice that number.
428571	thrice that number.
571428	four times that number.
714285	five times that number.
857142	six times that number.
999999	seven times that number.

Begin at the '1' in each line and it will be the same order of figures as the magic number up to six times that number, while seven times the magic number results in a row of 9's.

This is because the first six of these numbers appear in the cyclical decimal representation of the fraction

$$\frac{1}{7} = \frac{142857}{999999} = 0.142857142857142...$$

Another favourite multiplication trick, recalled by one of his child-friends, concerned the number 12 345 679 (there is no 8). At a children's party, Carroll asked a little boy to write it down:

He surveyed it in silence, then said, 'You don't form your figures very clearly, do you? Which of these figures do you think you have made the worst?' The boy thought his 5 was poorest. Lewis Carroll suggested he should multiply the line by 45. The child laboriously worked it out and to his surprise found the result was 555555555. 'Supposing I had said four, what then?' the boy queried. 'In that case we would have made the answer all fours,' Carroll replied. He would have told the boy to multiply by 36, another multiple of nine. But he did not attempt to explain 'mystic nines' to us.

The trick works because 9 × 12 345 679 = 111 111 111.

A well-known Victorian puzzle is now known as the *1089 puzzle*. Its originator is unknown, but may well have been Lewis Carroll.

Write down any three-digit number in which the first and last digits differ by more than 1. Reverse it, and subtract the smaller number from the larger. Reverse the answer and add the two last amounts together. The answer is always 1089, whatever number you started with.

For example, suppose that you write down	851
Reversing this gives	158
Subtracting the smaller amount from the larger gives	693
Reversing this gives	396
Adding these two last amounts gives	1089

Carroll certainly seems to have discovered a money version of this puzzle, involving pounds, shillings and pence. In pre-decimal coinage there were 12 pence in a shilling and 20 shillings in a pound.

> Put down any number of pounds not more than twelve, any number of shillings under twenty, and any number of pence under twelve. Under the pounds put the number of pence, under the shillings the number of shillings, and under the pence the number of pounds, thus reversing the line.
> Subtract.
> Reverse the line again.
> Add.
> Answer, £12 18s. 11d., *whatever* numbers may have been selected.

For example, suppose that you write down	£3 14s 9d.
Reversing this gives	£9 14s 3d.
Subtracting the smaller amount from the larger gives	£5 19s 6d.
Reversing this gives	£6 19s 5d.
Adding these last two amounts gives	£12 18s 11d.

Lewis Carroll's most ingenious money problem is a good test of our ability to work with pounds, shillings and pence, and involves a half-sovereign (10 shillings), a crown (5 shillings), a double-florin (4 shillings), a half-crown (2s. 6d.) and a florin (2 shillings). Since the double-florin was in circulation for only about four years, around 1890, the problem probably dates from this time.

> A customer bought goods in a shop to the amount of 7s. 3d. The only money he had was a half-sovereign, a florin, and a sixpence: so he wanted change. The shopman only had a crown, a shilling, and a penny. But a friend happened to come in, who had a double-florin, a half-crown, a fourpenny-bit, and a threepenny-bit.
> Could they manage it?

Another well-known puzzle that Carroll posed involves the mixing of brandy and water. The solution can be obtained by arithmetical calculation, but is more easily discovered by common-sense reasoning.

> Take two tumblers, one of which contains 50 spoonfuls of pure brandy and the other 50 spoonfuls of pure water. Take from the first of these one spoonful of the brandy and transfer it without spilling into the second tumbler and stir it up. Then take a spoonful of the mixture and transfer it back without spilling to the first tumbler.
>
> My question is, if you consider the whole transaction, has more brandy been transferred from the first tumbler to the second, or more water from the second tumbler to the first?

Geometrical Puzzles

Many of Lewis Carroll's puzzles have a geometrical flavour. In a letter to his child-friend Helen Feilden, Carroll posed his problem of the square window:

> A gentleman (a nobleman let us say, to make it more interesting) had a sitting-room with only one window in it — a square window, 3 feet high and 3 feet wide. Now he had weak eyes, and the window gave too much light, so (don't you like "so" in a story?) he sent for the builder, and told him to alter it, so as to give half the light. Only, he was to keep it square — he was to keep it 3 feet high — and he was to keep it 3 feet wide. How did he do it? Remember, he wasn't allowed to use curtains, or shutters, or coloured glass, or anything of that sort.

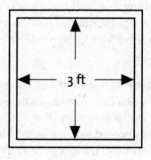

Another puzzle involving squares became popular in Victorian times, and was found among Carroll's papers after his death.

Start with an 8 × 8 grid of 64 squares and cut it into four pieces, as shown. If we rearrange the pieces, we obtain a 5 × 13 grid of 65 squares. Where did the extra square come from?

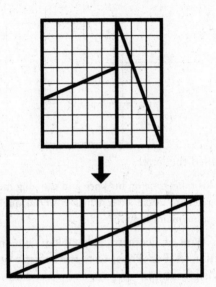

The numbers 5, 8 and 13 appearing here are all so-called Fibonacci numbers, a sequence that begins

1, 1, 2, 3, 5, 8, 13, 21, 34, 55, 89, ...

where each number after the first is the sum of the previous two. Carroll showed how to extend this seeming paradox to grids of squares involving larger Fibonacci numbers.

In August 1869 Carroll wrote to his child-friend Isabel Standen, mentioning another favourite puzzle:

Have you succeeded in drawing the three squares?

The challenge was to draw the following diagram without lifting one's pencil off the paper, repeating any line, or crossing over any existing line:

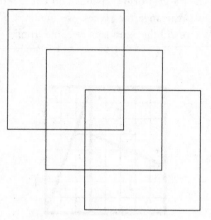

Later she recalled the letter:

> I remember his taking me on his knee and showing me puzzles, one of which he refers to in the letter, which is so thoroughly characteristic of him in its quaint humour.

A different type of geometrical problem, which Carroll mentioned in a letter to 'Bat' Price, tests the solver's powers of three-dimensional visualization:

> Imagine that you have some wooden cubes. You also have six paint tins each containing a different colour of paint. You paint a cube using a different colour for each of the six faces. How many different cubes can be painted using the same set of six colours? (Remember that two cubes are different only when it is not possible, by turning one, to make it correspond with the other.)

Another problem that Carroll sent to Bat Price involved symmetry, and is about clocks that look the same in our world and in looking-glass world:

> A clock face has all the hours indicated by the same mark, and both hands the same in length. It is opposite to a looking-glass. Find the time between 6 and 7 when the time as read direct and in the looking-glass shall be the same.

River-crossing Puzzles

River-crossing puzzles can be traced back as least as far as the ninth century, and take various forms. A typical example involves a fox, a goose and a bag of corn that a man was bringing back from market, as Carroll explained in a letter to his child-friend Jessie Sinclair:

> He had to get them over a river, and the boat was so tiny that he could only take *one* across at a time; and he couldn't ever leave the fox and the goose together, for then the fox would eat the goose; and if he left the goose and the corn together, the goose would eat the corn. So the only things he *could* leave safely together were the fox and the corn, for you never see a fox eating corn, and you hardly ever see a corn eating a fox.

A solution is as follows:

1. The man crosses the river with the goose, leaves it there, and returns alone.
2. He takes the corn across, leaves it there, and returns with the goose.
3. He takes the fox across, leaves him there with the corn, and returns alone.
4. He takes the goose across; all three are now on the other side.

Unfortunately, as Carroll later confessed, this puzzle was not always well received by his young friends:

> I must tell you an awful story of my trying to set a puzzle to a little girl the other day. It was at a dinner-party, at dessert. I had never seen her before, but, as she was sitting next me, I rashly proposed to her to try the puzzle (I daresay you know it) of 'the fox, and goose, and bag of corn.' And I got some biscuits to represent the fox and the other things. Her mother was sitting on the other side, and said, 'Now mind you take pains, my dear, and do it right!' The consequences were awful! She *shrieked* out "I can't do it! I can't do it! Oh, Mamma! Mamma!" threw herself into her mother's lap, and went off into a fit of sobbing which lasted several minutes! That was a lesson to me about trying children with puzzles.

On another occasion, Carroll called on the actress Ellen Terry at her home in Earl's Court, London, and presented a variation on the river-crossing puzzle to her son, the actor and designer Edward Gordon Craig, who recalled:

I can see him now, on one side of the heavy mahogany table — dressed in black, with a face which made no impression on me at all. I on the other side of the heavy mahogany table, and he describing in detail an event in which I had not the slightest interest — 'How five sheep were taken across a river in one boat, two each time — first two, second two — that leaves one yet two must go over' — ah — he did this with matches and a matchbox — I was not amused — so I have forgotten how these sheep did their trick.

A further variation of the river-crossing problem, devised by Carroll, involved a Queen who was imprisoned with her son and daughter at the top of a high tower.

Outside their window was a pulley with a rope round it, and a basket fastened at each end of the rope of equal weight. They managed to escape with the help of this and a weight they found in the room, quite safely. It would have been dangerous for any of them to come down if they weighed more than 15 lbs. more than the contents of the lower basket, for they would do so too quick, and they also managed not to weigh less either. The one basket coming down would naturally of course draw the other up. How did they do it? The Queen weighed 195 lbs., daughter 165, son 90, and the weight 75.

In an extension of this problem, also to be rescued were the Queen's pig (weighing 60 pounds), her dog (45 pounds) and her cat (30 pounds).

Other Recreations

We conclude this chapter with a miscellany of puzzles and problems that do not fit into the above categories.

The Number 42
Carroll seems to have had an obsession with the number 42. In *Alice's Adventures in Wonderland*, which has 42 illustrations, the number features in the Court scene:

At this moment the King, who had been for some time busily writing in his note-book, called out "Silence!" and read out from his book, "Rule Forty-two. *All persons more than a mile high to leave the court.*"

In *The Hunting of the Snark*, which Carroll started to write at the age of forty-two, the Preface mentions

Rule 42 of the Code, *"No one shall speak to the man at the helm"*

while the Baker managed to lose all of his luggage:

> He had forty-two boxes, all carefully packed,
> With his name painted clearly on each:
> But, since he omitted to mention the fact,
> They were all left behind on the beach.

And in an earlier poem, called *Phantasmagoria*, Carroll wrote:

> "No doubt," said I, "they settled who
> Was fittest to be sent:
> Yet still to choose a brat like you,
> To haunt a man of forty-two,
> Was no great compliment!"

More intriguing is Alice's comment after falling down the rabbit-hole:

> I'll try if I know all the things I used to know. Let me see: four times five is twelve, and four times six is thirteen, and four times seven is — oh dear! I shall never get to twenty at that rate! However, the Multiplication Table doesn't signify . . .

These calculations have been explained by means of different number bases. In everyday arithmetic we use numbers in base 10, where '42' means $(4 \times 10) + 2$. However, when counting shillings and pence or calculating with feet and inches, we use a number system based on 12: for example,

9d. + 5d. = 1s. 2d. and 9 in., + 5 in. = 1 ft. 2 in., so '9' + '5' = '12'

Returning to Alice's calculations, we note that

in base 18 arithmetic, $4 \times 5 = 20 = (1 \times 18) + 2$, which we write as '12'
in base 21 arithmetic, $4 \times 6 = 24 = (1 \times 21) + 3$, which we write as '13'

Continuing in this way, we have:

in base 24 arithmetic, $4 \times 7 = 28 = (1 \times 24) + 4$, which we write as '14'
in base 27 arithmetic, $4 \times 8 = 32 = (1 \times 27) + 5$, which we write as '15'

in base 30 arithmetic, $4 \times 9 = 36 = (1 \times 30) + 6$, which we write as '16'

in base 33 arithmetic, $4 \times 10 = 40 = (1 \times 33) + 7$, which we write as '17'

in base 36 arithmetic, $4 \times 11 = 44 = (1 \times 36) + 8$, which we write as '18'

in base 39 arithmetic, $4 \times 12 = 48 = (1 \times 39) + 9$, which we write as '19'

But now things go wrong:

in base 42 arithmetic, $4 \times 13 = 52 = (1 \times 42) + 10$, which we write as '1X'

where X is the symbol for 10 in base 42 — we do not get '20', which in base 42 corresponds to $(2 \times 42) + 0 = 84$.

So Alice was right!

Finding the Day of the Week

In March 1887 Lewis Carroll devised a method for mentally computing the day of the week for any given date (see opposite). Carroll could generally carry out these computations in his head in about 20 seconds.

Colouring Maps

A well-known problem in mathematics is the *four-colour problem*, first posed in 1852, which asks whether every map can be coloured with only four colours so that all regions sharing a border are coloured differently:

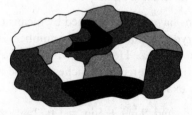

Carroll adapted it as a game for two people:

A is to draw a fictitious map divided into counties.

B is to colour it (or rather mark the counties with names of colours) using as few colours as possible.

Two adjacent counties must have different colours.

A's object is to force B to use as many colours as possible. How many can he force B to use?

Carroll's method for finding the day of the week for any given date

Take the given date in 4 portions, viz. the number of centuries, the number of years over, the month, the day of the month.

Compute the following 4 items, adding each, whenever found, to the total of the previous items. When an item or total exceeds 7, divide by 7, and keep the remainder only.

The Century-Item: Divide by 4, take overplus from 3, multiply remainder by 2.

The Year-Item: Add together the number of dozens, the overplus, and the number of 4's in the overplus.

The Month-Item: If it begins or ends with a vowel, subtract the number, denoting its place in the year, from 10. This, plus its number of days, gives the item for the following month. The item for January is "0"; for February or March (the third month), "3"; for December (the 12th month), "12."

The Day-Item: is the day of the month.

The total, thus reached, must be corrected, by deducting "1" (first adding 7, if the total be "0"), if the date be January or February in a Leap Year: remembering that every year, divisible by 4, is a Leap Year, excepting only the century-years, in New Style, when the number of centuries is *not* so divisible (*e.g.* 1800).

The final result gives the day of the week, "0" meaning Sunday, "1" Monday, and so on.

EXAMPLE
1783, September 18

17, divided by 4, leaves "1" over; 1 from 3 gives "2"; twice 2 is "4."

83 is 6 dozen and 11, giving 17; plus 2 gives 19, *i.e.* (dividing by 7) "5." Total 9, i.e. "2."

The item for August is "8 from 10," i.e. "2"; so for September, it is "2 plus 3," *i.e.* "5." Total 7, i.e. "0," which goes out.

18 gives "4." Answer, "*Thursday.*"

The Monkey and the Weight

A much discussed puzzle in Carroll's day concerned a monkey and a weight:

> A rope is supposed to be hung over a wheel fixed to the roof of a building; at one end of the rope a weight is fixed, which exactly counterbalances a monkey which is hanging on the other end. Suppose that the monkey begins to climb the rope, what will be the result?

In his diary entry for 21 December 1893, Carroll reported that he

> Got Prof. Clifton's answer to the "Monkey and Weight" problem. It is very curious, the different views taken by good mathematicians. Price says the weight goes *up*, with increasing velocity. Clifton (and Harcourt) that it goes *up*, as the same rate as the monkey, while Sampson says that it goes *down*!

and in a letter to Mrs Price, he confessed:

> It is a *most* puzzling puzzle . . . *my* present inclination is to believe that it goes *neither up nor down*!!!

Every Triangle is Isosceles

One geometrical paradox with which Carroll has been credited is the following 'proof' that *every triangle is isosceles* — that is, two of its sides have the same length. Can you find the error?

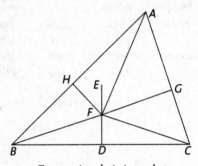

Every triangle is isosceles

Let *ABC* be any Triangle. Bisect *BC* at *D*, and from *D* draw *DE* at right angles to *BC*. Bisect the angle *BAC*.

(1) If the bisector does not meet *DE*, they are parallel. Therefore the bisector is at right angles to *BC*. Therefore, *AB* = *AC*, *i.e.*, *ABC* is isosceles.

(2) If the bisector meets *DE*, let them meet at *F*. Join *FB*, *FC*, and from *F* draw *FG*, *FH*, at right angles to *AC*, *AB*.

Then the Triangles *AFG*, *AFH* are equal, because they have the side *AF* common, and the angles *FAG*, *AGF* equal to *FAH*, *AHF*. Therefore *AH* = *AG*, and *FH* = *FG*.

Again, the Triangles *BDF*, *CDF* are equal, because *BD* = *DC*, *DF* is common, and the angles at *D* are equal. Therefore, *FB* = *FC*.

Again, the Triangles *FHB*, *FGC* are right-angled. Therefore the square on *FB* = the squares on *FH*, *HB*; and the square on *FC* = the squares on *FG*, *GC*. But *FB* = *FC*, and *FH* = *FG*. Therefore the square on *HB* = the square on *GC*. Therefore *HB* = *GC*. Also, *AH* has been proved = to *AG*. Therefore *AB* = *AC*, *i.e.*, *ABC* is isosceles. Therefore the Triangle *ABC* is always isosceles.

Q.E.D.

A Symmetric Poem

We conclude this Fit with an ingenious poem that has been attributed to Lewis Carroll. The words can be read either horizontally or vertically.

I	often	wondered	when	I	cursed,
Often	feared	where	I	would	be —
Wondered	where	she'd	yield	her	love
When	I	yield,	so	will	she,
I	would	her	will	be	pitied!
Cursed	be	love!	She	pitied	me . . .

Fit the Eighth
That's Logic

Alice meets Tweedledum and Tweedledee

Tweedledum: I know what you're thinking about, but it isn't so, nohow.

Tweedledee: Contrariwise, if it was so, it might be; and if it were so, it would be: but as it isn't, it ain't. That's logic.

As we saw in the Introduction, logical allusions and absurdities appear throughout Dodgson's books for children. His interest in logic dated from his undergraduate days when he was required to sit a logic paper as part of his Classical examinations. The subject permeated all aspects of his life, including religious matters, and in 1891, 'an old uncle, who has studied Logic for forty years' wrote to his nephew, Stuart Dodgson Collingwood, who was reading for Holy Orders:

> The bad logic that occurs in many and many a well-meant sermon, is a real danger to modern Christianity. When detected, it may seriously injure many believers, and fill them with miserable doubts. So my advice to *you*, as a young theological student, is "Sift your reasons *well*, and, before you offer them to others, make sure that they prove your conclusions."

Prim Misses and Sillygisms

References to logic in Dodgson's diaries occur as early as 1855. On 6 September of that year, he records that he

> Wrote part of a treatise on Logic, for the benefit of Margaret and Annie Wilcox.

Later, from around 1885 until his untimely death in 1898, he spent much of his time presenting symbolic logic as an entertainment for children to develop their powers of logical thought, and as a serious topic of study for adults. In order to increase their circulation, he published all these writings under his pen-name of Lewis Carroll.

As we shall see, much of his early work on logic was concerned with *syllogisms*, consisting of a couple of statements called *premisses* (now usually spelt 'premises') that lead to a *conclusion*. In *Sylvie and Bruno*, Arthur tries to explain these words to Lady Muriel:

> **Arthur:** For a *complete* logical argument we need two prim Misses —
> **Lady Muriel:** Of course! I remember that word now. And they produce — ?
> **Arthur:** A Delusion.
> **Lady Muriel:** Ye — es? I don't seem to remember that so well. But what is the *whole* argument called?
> **Arthur:** A Sillygism.

Syllogisms can be traced back to the fourth century BC, when Aristotle presented the following well-known example. Given the two premises

> All men are mortal
> All Greeks are men

we easily conclude that
> *All Greeks are mortal*

Less obvious is one of Carroll's syllogisms. Starting with the premises

> Some new cakes are unwholesome
> No nice cakes are unwholesome

Carroll concluded that
> *Some new cakes are not nice*

In the next section we explain how Carroll reached this conclusion.

Note that in each of these syllogisms, the two premises have a common term (*men* and *unwholesome*) that does not feature in the conclusion. To find this conclusion, we need to eliminate this common term.

Carroll constructed many syllogisms in which one needs to do this. Among the more entertaining are the following, which may also be worked out using the method of the next section:

> No fossil can be crossed in love
> An oyster may be crossed in love

Conclusion:
> *Oysters are not fossils*

> A prudent man shuns hyaenas
> No banker is imprudent

Conclusion:
> *No banker fails to shun hyaenas*

> No bald creature needs a hairbrush
> No lizards have hair

Conclusion:
> *No lizard needs a hairbrush*

But there is no reason why we should restrict ourselves to only two premises. One of Carroll's simpler examples featured three:

Babies are illogical
Nobody is despised who can manage a crocodile
Illogical persons are despised

We can easily sort these out. From the first and third premises we can eliminate the common term *illogical* and conclude that
Babies are despised

But the second premise tells us that despised persons cannot manage crocodiles. We can therefore eliminate the common term *despised* and conclude that
Babies cannot manage crocodiles

This example yielded two syllogisms — one consisting of the first and third premises, and the other combining their conclusion with the second premise. From each syllogism we have eliminated the common term (*illogical* or *despised*) and reached a conclusion that involves only the remaining terms (*babies* and *managing crocodiles*).

Such a succession of premises that can be simplified two at a time to yield an eventual conclusion is called a *sorites* (pronounced 'so-right-ease'). Lewis Carroll constructed a large number of them. Here is one with four premises:

No birds, except ostriches, are 9 feet high
There are no birds in this aviary that belong to any one but me
No ostrich lives on mince-pies
I have no birds less than 9 feet high

Using all this information, he was able to conclude that
No bird in this aviary lives on mince-pies

Another sorites of Carroll's has five premises:

No kitten, that loves fish, is unteachable
No kitten without a tail will play with a gorilla
Kittens with whiskers always love fish
No teachable kitten has green eyes
No kittens have tails unless they have whiskers

from which he concluded that
No kitten with green eyes will play with a gorilla

But these are far from obvious. How might we go about sorting them out? Carroll used a systematic pictorial method that we now describe.

The Game of Logic

From now on, any statement, whether a premise, a conclusion, or any other assertion, will be called a *proposition*. The types of proposition that Lewis Carroll was dealing with are those that can be written in one of three forms, as illustrated by the following sentences:

Some new cakes are nice
No new cakes are nice
All new cakes are nice

For example, he would write

I have no birds less than 9 feet high

in the second of these forms:

No birds that belong to me are less than 9 feet high

and he would write

Kittens with whiskers always love fish

in the third form:

All kittens that have whiskers love fish

He also gave examples of more complicated sentences that can be written in one of these forms; for example, he expressed the proposition

None but the brave deserve the fair

in the form

No not-brave persons are persons that deserve the fair

and he rewrote

Happy is the man who does not know what 'toothache' means!

in the form

All men who do not know what 'toothache' means are happy men

The reason for doing this was so that he could work exclusively with these three types of proposition, which we may express in symbolic form as

Some x are y No x are y All x are y

The objects x and y should be of the same type, such as *birds*, *cakes*, *persons* or *puppies*; in the language of logic we call such a collection the *universe*.

When explaining syllogisms, Carroll found it helpful to introduce pictures to represent propositions. Each of these pictures, which he called *biliteral diagrams*, has the form of a square representing the universe. As he remarked to his young readers,

"Let us take a Universe of Cakes." (Sounds nice, doesn't it?)

This square is then divided into four smaller squares. In the top row are the objects x (such as *new cakes*), and in the bottom row the objects that are not x (*not-new cakes*). Similarly, the left column contains the objects y (such as *nice cakes*) and the right column the objects that are not y (*not-nice cakes*). Each of the four smaller squares now refers to a certain collection of cakes — for example at the top right are the *new not-nice cakes*.

	y	not-y		nice cakes	not-nice cakes
x	x and y	x, not-y	new cakes	new nice cakes	new not-nice cakes
not-x	y, not-x	not-x, not-y	not-new cakes	not-new nice cakes	not-new not-nice cakes

Biliteral diagrams

In order to represent propositions on these diagrams, Carroll used a red counter, denoted here by ●, to indicate the presence of an object, and a grey one, denoted by ○, to indicate its absence. For example, he illustrated the statement 'Some new cakes are nice' by placing a red counter in the top-left corner, and the statement 'No new cakes are nice' by placing a grey counter there. More generally, we can illustrate such propositions about objects x and y in a similar way.

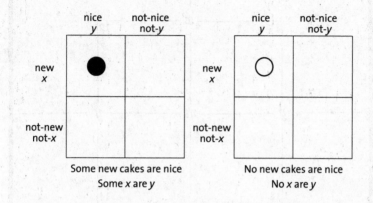

Some new cakes are nice
Some x are y

No new cakes are nice
No x are y

To illustrate the statement 'All new cakes are nice', Carroll first replaced it by the two equivalent propositions 'Some new cakes are nice' and 'No new cakes are not-nice', and then placed two counters in the appropriate corners:

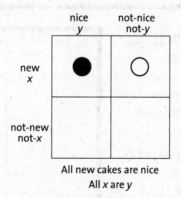

All new cakes are nice
All x are y

For some time, Carroll had nursed the idea of writing a multi-volume work on symbolic logic, or *Logic for Ladies* as he originally intended to call it, but on 24 July 1886 he had a change of plan:

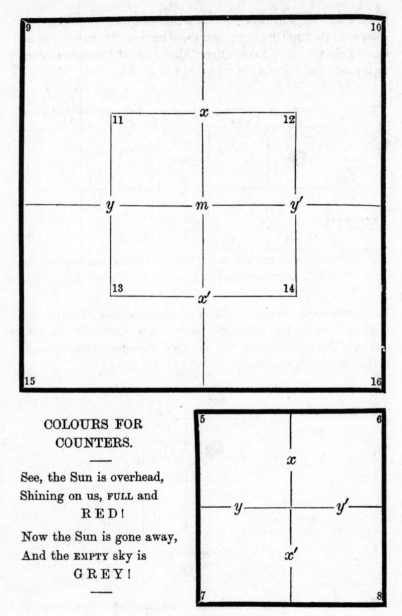

COLOURS FOR
COUNTERS.

—

See, the Sun is overhead,
Shining on us, FULL and
RED!

Now the Sun is gone away,
And the EMPTY sky is
GREY!

—

The board for The Game of Logic

The idea occurred to me this morning of beginning my "Logic" publication, not with "Book I" of the full work "Logic for Ladies," but with a small pamphlet and a cardboard diagram, to be called The Game of Logic. I have during the day written most of the pamphlet.

The Game of Logic was designed to convey the ideas of syllogisms to young people. By the end of the year his 'small pamphlet' had expanded to a book of about a hundred pages, consisting of four parts:

New Lamps for Old, presenting his pictorial approach to propositions and syllogisms;
Cross Questions, containing a number of problems to be solved;
Crooked Answers, giving their solutions;
Hit or Miss, containing a large number of unsolved problems.

It came with a board and nine counters, four red ones and five grey ones.

As Carroll observed in the book's preface, a game needs players:

Besides the nine Counters, it also requires one Player, at least. I am not aware of any Game that can be played with less than this number: while there are several that require more: take Cricket, for instance, which requires twenty-two. How much easier it is, when you want to play a Game, to find one Player than twenty-two. At the same time, though one Player is enough, a good deal more amusement may be got by two working at it together, and correcting each other's mistakes . . . it will give the Players a little instruction as well. But is there any great harm in that, so long as you get plenty of amusement?

We can illustrate Carroll's method for sorting out syllogisms by working through his first example:

Some new cakes are unwholesome
No nice cakes are unwholesome

Here, the cakes are described in three ways as new, nice and wholesome, so we need to divide up the board in three ways, giving a triliteral diagram:

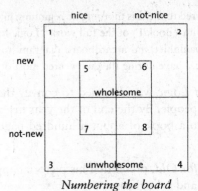

Numbering the board

the top half represents *new* and the bottom half *not-new*;
the left half represents *nice* and the right half *not-nice*;
the centre represents *wholesome* and the rest *unwholesome*.

Notice that the common description *wholesome* appears in the centre. So, if we number the sections of the board as shown, then *new* corresponds to the upper regions, 1, 2, 5 and 6; *nice* corresponds to the left-hand regions, 1, 3, 5 and 7; and *wholesome* corresponds to the central regions, 5, 6, 7 and 8. For example, region 6 contains the *new not-nice wholesome* cakes, and region 4 contains the *not-new not-nice unwholesome* cakes ('Remarkably untempting Cakes!', as Carroll observed).

The object is to eliminate the centre and obtain a biliteral diagram representing the conclusion. To do this, we look at the propositions one at a time, starting with any negative ones. In general, we consider negative statements first, since they enable us to place grey counters with certainty. The second proposition,

No nice cakes are unwholesome

tells us that none of the cakes in the left half (the *nice* cakes) lies in the centre (the *wholesome* cakes). This means that regions 1 and 3 are empty, and we indicate this by placing grey counters in these regions:

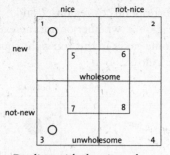

Dealing with the nice cakes

Next we consider the first proposition,

Some new cakes are unwholesome

This tells us that some of the cakes in the top half (the *new* cakes) do not lie in the centre. We can indicate this by placing a red counter in region 1 or region 2. But region 1 already contains a grey counter (there are no cakes there), so we must place our red counter in region 2:

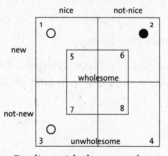

Dealing with the new cakes

Finally, we transfer the information to the biliteral diagram at the bottom of the board in order to obtain a connection between the *new* cakes and the *nice* ones. The top-left square corresponds to regions 1 and 5 above; we know that region 1 is empty, but we can say nothing about region 5 — so we do not know anything with certainty about this top-left square. Similarly, we do not know anything with certainty about the bottom-left or the bottom-right square. But the top-right square corresponds to regions 2 and 6, and we know that some cakes lie in region 2 (whether or not there are any in region 6), so there must be some cakes in this top-right square.

The result is as follows:

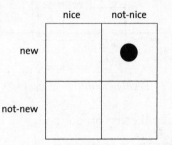

Transferring the information to the smaller square

The conclusion is that

Some new cakes are not nice

There is one further comment to make. Sometimes we do not know on which of two adjacent regions a counter should be placed; in such a case, we place it on the boundary line between them. For example, consider the single proposition

Some wholesome cakes are new

Here the red counter must be placed in region 5 or region 6, but we do not know which, so we place it as follows until we have more information:

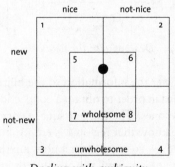

Dealing with ambiguity

Carroll described this situation as follows:

> Our ingenious American cousins have invented a phrase to express the position of a man who wants to join one or other of two parties — such as their two parties 'Democrats' and 'Republicans' — but ca'n't make up his mind which. Such a man is said to be "sitting on the fence." Now that is exactly the position of the red counter you have just placed on the division-line. He likes the look of No. 5, and he likes the look of No. 6, and he doesn't know which to jump down into. So there he sits astride, silly fellow, dangling his legs, one on each side of the fence!

After a bit of practice, the above method becomes quite simple to apply. However, we can shorten the discussion by introducing some symbols. Let us use the letters

x = new cakes, y = nice cakes, z = wholesome cakes

and denote their opposites with dashes:

x' = not-new cakes, y' = not-nice cakes,
z' = unwholesome cakes

Then our syllogism can be written more concisely as

Some x are z' & No y are z' Some x are y'

Symbolic Logic

Carroll was so keen to introduce young people to the delights of symbolic logic that for several years he presented classes on logic at the Oxford High School for Girls and in two of the Oxford Colleges, Lady Margaret Hall and St Hugh's Hall. As he wrote in a letter to a friend:

> *Every* afternoon, oddly enough, I have an engagement, as I have taken to giving lectures, on my *Game of Logic*, to young people ... Girls are *very* nice pupils to lecture to, they are so bright and eager.

His enthusiasm was sometimes reciprocated. One of his pupils recalls how on the first occasion of his coming to hold a Logic class at St. Hugh's, about a dozen students assembled solemnly in the library, armed with note-books and pencils, prepared to listen to a serious lecture on a difficult subject. To their surprise, and also somewhat to their dismay, Mr. Dodgson produced from his black bag twelve large white envelopes, each containing a card marked with a diagram, and a set of counters in two colours. These he dealt out to his audience. "Now," he said cheerfully, "I will teach you to play the game of Logic!" And then, when he proceeded to illustrate his explanations with examples, his pupils found that they were actually expected to *laugh*! But though such propositions as:

> "Some new Cakes are nice"
> "No new Cakes are nice"
> "All new Cakes are nice"

and

> "All teetotallers like sugar"
> "No nightingale drinks wine,"

sounded rather like extracts from a child's reading-book, it was soon discovered that considerable intelligence, as well as much skill and attention were required to learn the game and work out a conclusion on the diagram. How patiently he bore with our stupidity! To him we were all still very young!

However, not everyone was so entranced, as one child-friend who had spent a holiday in Eastbourne with Carroll later recalled:

His great delight was to teach me his Game of Logic. Dare I say this made the evening rather long, when the band was playing outside on the parade and the moon shining on the sea.

and The *Cambridge Review* was also unenthusiastic:

The author has attempted to enliven the subject with that playful humour which made "Alice in Wonderland" so delightful; but in this case the logic seems to overpower the humour, so that we can with some confidence recommend the book to those suffering from insomnia. To make the book palatable to children, for whom it is apparently written, frequent practical illustrations of the introductory proposition, "Some new cakes are nice", will be required.

On 21 August 1894 Charles Dodgson wrote to Mary Brown, a child-friend from many years earlier, about his major new project:

> At present I'm hard at work (and have been for months) on my Logic-book. (It has really been on hand for a dozen years: the "months" refer to preparing for the Press.) It is *Symbolic Logic*, in 3 parts — and Part I is to be easy enough for boys and girls of (say) 12 or 14. I greatly hope it will get into High Schools, etc. I've been teaching it at Oxford to a class of girls at the High School, another class of the mistresses (!), and another class of girls at one of the Ladies' Colleges. I believe it's one of the *best* mental exercises that the young could have: and it doesn't need *special* powers like mathematics. I may *perhaps* get Part I out early next year. The next will take another year at least. I think I once gave you my *Game of Logic*? This is a more serious attempt: but with much shorter (and, I hope, better) explanations.

The first edition of *Symbolic Logic*, Part I: *Elementary* appeared in February 1896, and the 500 copies, costing 2 shillings, sold out immediately. Promoted as 'A fascinating mental recreation for the young', the work was 'Dedicated to the memory of Aristotle' and opens with some encouragement for Carroll's young readers:

> If, dear Reader, you will faithfully observe these Rules, and so give my book a really *fair* trial, I promise you, most confidently, that you will find Symbolic Logic to be one of the most, if not *the* most, fascinating of mental recreations! . . . I have myself taught most of its contents, *vivâ voce*, to *many* children, and have found them take a real intelligent interest in the subject.

Elsewhere, the author writes:

> It may *look* very difficult, at first sight; but I have taught it, with ease, to many children: my typical case being a little niece of my own, aged 9, who took in nearly the whole of the book in 4 lessons, and seemed to enjoy it thoroughly — specially the enunciation, at all convenient opportunities, of the sonorous phrase "Dichotomy by Contradiction"!

Most of Part I is concerned with the sorting-out of syllogisms by means of his biliteral and triliteral diagrams, and with the

A Syllogism worked out.

That story of yours, about your once meeting the sea-serpent, always sets me off yawning;

I never yawn, unless when I'm listening to something totally devoid of interest.

The Premisses, separately.

The Premisses, combined.

The Conclusion.

That story of yours, about your once meeting the sea-serpent, is totally devoid of interest.

The frontispiece to Symbolic Logic, *Part I*

extension of these ideas to a sorites with more than two propositions. The language he uses is more sophisticated than in *The Game of Logic*, and he employs notation to shorten the exposition. For example, starting with the propositions

> Babies are illogical
> Nobody is despised who can manage a crocodile
> Illogical persons are despised

he writes

> Univ. "persons"; a = able to manage a crocodile; b = babies;
> c = despised; d = logical

and uses dashes for their opposites, so that the propositions become

> All b are d≢   No a are c and   All d≢ are c

Using the above method to eliminate the common term d from the first and third propositions yields

> All b are c

Combining this with the second proposition then yields the conclusion

> All b are a≢: *Babies cannot manage crocodiles*

Pairs of Premisses for Syllogisms. (*Answers only are supplied.*)
———

1. All pigs are fat;
   Nothing that is fed on barley-water is fat.
2. All rabbits, that are not greedy, are black;
   No old rabbits are free from greediness.
3. Some pictures are not first attempts;
   No first attempts are really good.
4. Toothache is never pleasant;
   Warmth is never unpleasant.
5. I never neglect important business;
   Your business is unimportant.
6. No pokers are soft;
   All pillows are soft.
7. Some lessons are difficult;
   What is difficult needs attention.
8. All clever people are popular;
   All obliging people are popular.
9. Thoughtless people do mischief;
   No thoughtful person forgets a promise.
10. Pigs cannot fly;
    Pigs are greedy.

A proof copy of Carroll's syllogisms, with his notation for solving them

For a more complicated example, we return to our earlier example with four propositions:

No birds, except ostriches, are 9 feet high
There are no birds in this aviary that belong to any one but me
No ostrich lives on mince-pies
I have no birds less than 9 feet high

Here Carroll writes:

Univ. "birds"; a = in this aviary; b = living on mince-pies; c = my; d = 9 feet high; e = ostriches

so that the propositions become

No e' are d, No c' are a, No e are b, No d' are c

We now look for common terms. Eliminating the common term e from the first and third propositions yields

No b are d

Next, combining this with the fourth proposition yields

No b are c

Finally, combining this with the second proposition yields the conclusion:

No a are b: No bird in this aviary lives on mince-pies

The final section of *Symbolic Logic*, Part I contains no fewer than sixty such examples, with up to ten propositions. In each case we can apply the same method, choosing pairs of propositions with a common term and eliminating these one by one.

In an 'Appendix, Addressed to Teachers' at the end of Part I, Carroll discusses various subjects in greater depth, and goes on to outline some of the more advanced topics he was proposing to deal with in Parts II and III. Among these are several more complicated examples with up to twenty propositions.

For those who succeed in mastering Part I, and who begin, like Oliver, "asking for more," I hope to provide, in Part II, some

The Problem of the School-Boys

All the boys, in a certain School, sit together in one large room every evening. They are of no less than *five* nationalities — English, Scotch, Welsh, Irish, and German. One of the Monitors (who is a great reader of Wilkie Collins' novels) is very observant, and takes MS. notes of almost everything that happens, with the view of being a good sensational witness, in case any conspiracy to commit a murder should be on foot. The following are some of his notes:—

(1) *Whenever some of the English boys are singing "Rule Britannia", and some not, some of the Monitors are wide-awake;*

(2) *Whenever some of the Scotch are dancing reels, and some of the Irish fighting, some of the Welsh are eating toasted cheese;*

(3) *Whenever all the Germans are playing chess, some of the Eleven are not oiling their bats;*

(4) *Whenever some of the Monitors are asleep, and some not, some of the Irish are fighting;*

(5) *Whenever some of the Germans are playing chess, and none of the Scotch are dancing reels, some of the Welsh are not eating toasted cheese;*

(6) *Whenever some of the Scotch are not dancing reels, and some of the Irish not fighting, some of the Germans are playing chess;*

(7) *Whenever some of the Monitors are awake, and some of the Welsh are eating toasted cheese, none of the Scotch are dancing reels;*

(8) *Whenever some of the Germans are not playing chess, and some of the Welsh are not eating toasted cheese, none of the Irish are fighting;*

(9) *Whenever all the English are singing "Rule Britannia", and some of the Scotch are not dancing reels, none of the Germans are playing chess;*

(10) *Whenever some of the English are singing "Rule Britannia", and some of the Monitors are asleep, some of the Irish are not fighting;*

(11) *Whenever some of the Monitors are awake, and some of the Eleven are not oiling their bats, some of the Scotch are dancing reels;*

(12) *Whenever some of the English are singing "Rule Britannia", and some of the Scotch are not dancing reels,* * * * *

Here the MS. breaks off suddenly. The Problem is to complete the sentence, if possible.

tolerably hard nuts to crack — nuts that will require all the nut-crackers they happen to possess!

We give one of these, with twelve premises, on page 189. It can be solved using the ideas illustrated on the preceding pages.

Sadly, Charles Dodgson died before Part II of his *Symbolic Logic* was completed. He had prepared examples with as many as fifty propositions, a general method he devised for solving puzzles of this kind (the 'Method of Trees'), and a number of logic puzzles and paradoxes that we present below. Parts of the book were in proof form, but they disappeared and were not rediscovered until many years after his death.

Venn, Carroll and Churchill

Lewis Carroll was not the first to use pictorial methods to solve logic problems. In 1768 Leonhard Euler, the most prolific mathematician of all time, introduced some circle diagrams in his celebrated *Letters to a German Princess*. A much improved version was produced by the Cambridge mathematician John Venn in 1880, and such diagrams are now usually known as *Venn diagrams*.

Venn's two-circle version of a biliteral diagram is shown below. According to Carroll, Venn used a plus sign to indicate a region that is occupied, and shading for one that is empty. His diagrams did not include a universe (which is taken to be the exterior region), a fact greatly criticized by Carroll.

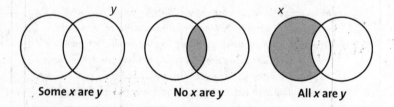

Some *x* are *y*   No *x* are *y*   All *x* are *y*

Venn's well-known three-circle diagrams can be used to represent syllogisms employing three letters, *x*, *y* and *z* (see opposite).

Here the eight regions (including the exterior) represent all possibilities for *x*, *y* and *z* and their opposites, *x*≇, *y*≇ and *z*≇ — for example, the shaded region represents the combination *x*, *y*

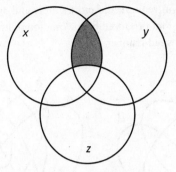

Venn's three-circle diagram

and $z \nleq$. Such a diagram was drawn by Winston Churchill at Hever Castle on 5 June 1948

to illustrate England's partition in the world-to-be "IF WE ARE WORTHY".

His circles show the mutual relationships between BE (the British Empire), UE (United Europe) and ESW (the English Speaking World of about 200 million people):

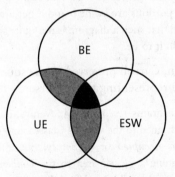

Churchill's three-circle diagram

The situation becomes more difficult when we have more than three letters, because it is no longer possible to represent them by circles. Venn's diagram for four letters is shown overleaf. As he remarked,

With four terms in request, the most simple and symmetrical solution seems to me that produced by making four ellipses intersect one another in the desired manner.

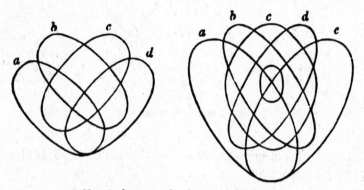

Venn's diagrams for four and five letters

For five letters, he produced a diagram with one portion split into two:

> The simplest diagram I can suggest is one like this (the small ellipse in the centre is to be regarded as a portion of the outside of c; i.e. its four component portions are inside b and d but are no part of c). It must be admitted that such a diagram is not quite so simple to draw as one might wish it to be.

As we have seen, Carroll preferred to use square drawings, with the outside square representing the universe:

> My method of diagrams *resembles* Mr. Venn's, in having separate Compartments assigned to the various Classes, and in marking these Compartments as *occupied* or as *empty*; but it *differs* from his Method, in assigning a *closed* area to the *Universe of Discourse*, so that the Class which, under Mr. Venn's liberal sway, has been ranging at will through Infinite Space, is suddenly dismayed to find itself "cabin'd, cribb'd, confined", in a limited Cell like any other Class!

For two and three letters, Carroll used his biliteral and triliteral diagrams. For four letters he replaced the central square by a vertical rectangle and added a horizontal one, and for five letters he divided each region into two halves. Venn's diagrams went up to only five letters, but Carroll continued further: his diagrams for six, seven and eight letters are as shown here, and he also described constructions for nine and ten letters and more.

Carroll's diagrams for four to eight letters

Logic Puzzles and Paradoxes

In the 1890s Carroll invented a number of ingenious logic puzzles and paradoxes that he planned to use in Part II of *Symbolic Logic*. He based some of these on ideas that have 'come down to us from ancient times', being developments of the ancient *liar paradox*, in which the sixth-century BC Cretan poet Epimenides asserted that

> Cretans always tell lies

The paradox arises from the fact that if Epimenides is telling the truth, then, being a Cretan, he must be a liar; if he is lying, then, being a Cretan, he must be telling the truth.

We present one of Carroll's adaptations of the liar paradox, to which he gave the classical title *Crocodilus*:

> A Crocodile had stolen a Baby off the banks of the Nile. The Mother implored him to restore her darling. "Well," said the Crocodile, "if you say truly what I shall do I will restore it: if not, I will devour it."

193

"You will devour it!" cried the distracted Mother. "Now," said the wily Crocodile, "I cannot restore your Baby: for if I do, I shall make you speak falsely: and I warned you that, if you spoke falsely, I would devour it." "On the contrary," said the yet wilier Mother, "you cannot devour my Baby: for if you do, you will make me speak truly, and you promised me that, if I spoke truly, you would restore it!" (We assume, of course, that he was a Crocodile of his word; and that his sense of honour outweighed his love of babies.)

Carroll's conclusion was that whatever the Crocodile does, he *breaks* his word.

Another logical paradox concerned heights:

> Men over 5 feet high are numerous
> Men over 10 feet high are not numerous

Conclusion:

> *Men over 10 feet high are not over 5 feet high*

Yet another concerned a small girl and her sympathetic friend:

> **Small Girl:** I'm *so* glad I don't like asparagus!
> **Friend:** Why, my dear?
> **Small girl:** Because, if I did, I should have to eat it — and I ca'n't bear it!

But probably the most controversial of his logical problems was *A Logical Paradox*, dating from 1892, which later appeared in an extended version in the periodical *Mind* (a quarterly review of psychology and philosophy) and also in several shorter forms. One of these concerned three barbers, Allen, Brown and Carr, who cannot all leave their barber's shop at the same time.

> Suppose that:
>> If Carr is out, then if Allen is out, Brown is in.
>> If Allen is out, Brown is out.
> Can Carr go out?

Oxford's Wykeham Professor of Logic, John Cook Wilson, became interested in this problem in December 1892, and spent the next few years disagreeing with Carroll about it. He argued that if Carr goes out, then the two conditions become contradictory — so Carr cannot go out. Dodgson agreed that Carr and Allen cannot go out together, since then Brown would have to be both in and out, but argued that Carr can certainly go out on his own as long as Allen stays in.

This dispute eventually went public, with Carroll producing pamphlets presenting both sides of the argument under the pseudonyms of 'Nemo' for Wilson and 'Outis' for himself; *nemo* and *outis* are the Latin and Greek words for 'nobody':

> Have just got printed, as a leaflet, *A Disputed Point in Logic*, the point Prof. Wilson and I have been arguing so long. The paper is *wholly* in his own words, and puts the point very clearly. I think of submitting it to all my logical friends.

Eventually Nemo gave in, but not until seven years after Outis's death.

What the Tortoise Said to Achilles

Another ancient mathematical paradox, dating from the fifth century BC, is known as *Zeno's paradox* and concerns a race between Achilles and a Tortoise.

Let us suppose that Achilles runs ten times as fast as the Tortoise, and that Achilles gives the Tortoise a 100-metre start before starting to catch up. By the time Achilles reaches the Tortoise's starting point, the Tortoise will have moved on by ten metres. By the time Achilles reaches this next point, the Tortoise will have moved on yet another metre, and so on, for ever. It follows that Achilles can never catch the Tortoise — and yet he clearly does. How can this be explained?

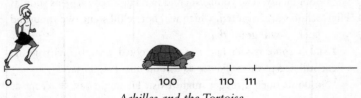

Achilles and the Tortoise

Dodgson frequently discussed this paradox; for example, a version of it appeared as Problem 2 of Knot IX in *A Tangled Tale* (see Fit the Seventh). It arises from the fact that we can add up *infinitely many* distances and yet the *total distance* may be *finite*. In this case, the total distance that Achilles must run to catch the Tortoise is

$$100 + 10 + 1 + \frac{1}{10} + \frac{1}{100} + \frac{1}{1000} + \ldots$$

which is 111.111... or $111\frac{1}{9}$.

In 1895, while working on some problems in logic, he published a sequel to it in *Mind*; it is couched in the language of Euclidean geometry (see Fit the Fourth). Bertrand Russell later described this and the barber's shop problem as Dodgson's greatest contributions to logic.

As we join Carroll in his exposition, Achilles has just overtaken the Tortoise and is sitting comfortably upon its back. After some initial banter the two characters begin a geometrical discussion:

"That beautiful First Proposition of Euclid!" the Tortoise murmured dreamily. "You admire Euclid?"

"Passionately! So far, at least, as one *can* admire a treatise that wo'n't be published for some centuries to come!"

"Well, now, let's take a little bit of the argument in that First Proposition — just *two* steps, and the conclusion drawn from them. Kindly enter them in your note-book. And in order to refer to them conveniently, let's call them *A*, *B*, and *Z*:–

(*A*) Things that are equal to the same are equal to each other.

(*B*) The two sides of this Triangle are things that are equal to the same.

(*Z*) The two sides of this Triangle are equal to each other.

"Readers of Euclid will grant, I suppose, that *Z* follows logically from *A* and *B*, so that any one who accepts *A* and *B* as true, *must* accept *Z* as true?"

"Undoubtedly! The youngest child in a High School — as soon as High Schools are invented, which will not be till some two thousand years later — will grant *that*."

"And if some reader had *not* yet accepted *A* and *B* as true, he might still accept the *sequence* as a *valid* one, I suppose?"

"No doubt such a reader might exist. He might say, 'I accept as true the Hypothetical Proposition that, *if A* and *B* be true, *Z* must be true; but I *don't* accept *A* and *B* as true.' Such a reader would do wisely in abandoning Euclid, and taking to football."

"And might there not *also* be some reader who would say 'I accept *A* and *B* as true, but I *don't* accept the Hypothetical'?"

"Certainly there might. *He*, also, had better take to football."

"And *neither* of these readers," the Tortoise continued, "is *as yet* under any logical necessity to accept Z as true?"

"Quite so," Achilles assented.

"Well, now, I want you to consider *me* as a reader of the *second* kind, and to force me, logically, to accept Z as true."

"A tortoise playing football would be —" Achilles was beginning

"— an anomaly, of course," the Tortoise hastily interrupted. "Don't wander from the point. Let's have Z first, and football afterwards!"

"I'm to force you to accept Z, am I?" Achilles said musingly. "And your present position is that you accept A and B, but you *don't* accept the Hypothetical —"

"Let's call it C," said the Tortoise.

"— but you *don't* accept:

(C) If A and B are true, Z must be true."

"That is my present position," said the Tortoise.

"Then I must ask you to accept C."

"I'll do so," said the Tortoise, "as soon as you've entered it in that note-book of yours. What else have you got in it?"

"Only a few memoranda," said Achilles, nervously fluttering the leaves: "a few memoranda of — of the battles in which I have distinguished myself!"

"Plenty of blank leaves, I see!" the Tortoise cheerily remarked. "We shall need them *all*!" (Achilles shuddered.) "Now write as I dictate:–

(A) Things that are equal to the same are equal to each other.

(B) The two sides of this Triangle are things that are equal to the same.

(C) If A and B are true, Z must be true.

(Z) The two sides of this Triangle are equal to each other."

"You should call it D, not Z," said Achilles. "It comes *next* to the other three. If you accept A and B and C, you *must* accept Z."

"And why *must* I?"

"Because it follows *logically* from them. If A and B and C are true, Z *must* be true. You can't dispute *that*, I imagine?"

"If A and B and C are true, Z *must* be true," the Tortoise thoughtfully repeated. "That's *another* Hypothetical, isn't it? And, if I failed to see its truth, I might accept A and B and C, and *still* not accept Z, mightn't I?"

"You might," the candid hero admitted; "though such obtuseness would certainly be phenomenal. Still, the event is *possible*. So I must ask you to grant *one* more Hypothetical."

"Very good, I'm quite willing to grant it, as soon as you've written it down. We will call it

(D) If A and B and C are true, Z must be true.

Have you entered that in your note-book?"

"I *have*!" Achilles joyfully exclaimed, as he ran the pencil into its sheath. "And at last we've got to the end of this ideal race-course! Now that you accept A and B and C and D, *of course* you accept Z."

"Do I?" said the Tortoise innocently. "Let's make that quite clear. I accept A and B and C and D. Suppose I *still* refused to accept Z?"

"Then Logic would take you by the throat, and *force* you to do it!" Achilles triumphantly replied. "Logic would tell you 'You ca'n't help yourself. Now that you've accepted A and B and C and D, you *must* accept Z.' So you've no choice, you see."

"Whatever *Logic* is good enough to tell me is worth *writing down*," said the Tortoise. "So enter it in your book, please. We will call it

(E) If A and B and C and D are true, Z must be true. Until I've granted *that*, of course I needn't grant Z. So it's quite a *necessary* step, you see?"

"I see," said Achilles; and there was a touch of sadness in his tone.

Here the narrator, having pressing business at the Bank, was obliged to leave the happy pair, and did not again pass the spot until some months afterwards. When he did so, Achilles was still seated on the back of the much-enduring Tortoise, and was writing in his note-book, which appeared to be nearly full. The Tortoise was saying, "Have you got that last step written down? Unless I've lost count, that makes a thousand and one. There are several millions more to come. And *would* you mind, as a personal favour, considering what a lot of instruction this colloquy of ours will provide for the Logicians of the Nineteenth Century — *would* you mind adopting a pun that my cousin the Mock-Turtle will then make, and allowing yourself to be re-named *Taught-Us*?"

"As you please!" replied the weary warrior, in the hollow tones of despair, as he buried his face in his hands. "Provided that *you*, for *your* part, will adopt a pun the Mock-Turtle never made, and allow yourself to be re-named *A Kill-Ease*!"

Conclusion
Math and Aftermath

Symbolic logic was not the only preoccupation of Charles Dodgson in the 1890s. In the last few years of his life he dabbled in a range of mathematical topics, a few of which we outline here.

Pillow-Problems

A substantial collection of problems, *Curiosa Mathematica*, Part II: *Pillow-Problems Thought out During Sleepless Nights*, appeared in 1893. In the Introduction the author describes how it came into being:

> Nearly all of the following seventy-two Problems are veritable "Pillow-Problems", having been solved, in the head, while lying awake at night . . . every one of them was worked out, to the very end, before drawing any diagram or writing down a single word of the solution. I generally wrote down the *answer*, first of all: and *afterwards* the question and its solution.

His reason for publishing these imaginative problems was not to show off his powers of mental calculation — which were truly remarkable — but rather to encourage other mathematicians to improve their own mental problem-solving abilities.

> My purpose — of giving this encouragement to others — would not be so well fulfilled had I allowed myself, in writing out my solutions, to *improve* on the work done in my head. I felt it to be much more important to set down *what had actually been done in the head*, than to supply shorter or neater solutions, which perhaps would be much harder to do without paper.

In the second edition Dodgson altered the wording of the subtitle to say *Wakeful Hours* rather than *Sleepless Nights*:

> This last change has been made in order to allay the anxiety of kind friends, who have written to me to express their sympathy in my broken-down state of health, believing that I am a sufferer from

Charles Dodgson

chronic "insomnia", and that it is as a remedy for that exhausting malady that I have recommended mathematical calculation.

Many of the Pillow-Problems are highly ingenious. They range across many areas of mathematics: arithmetic, algebra, pure and algebraic geometry, trigonometry, differential calculus and probabilities. A representative selection of them follows.

8. Some men sat in a circle, so that each had 2 neighbours; and each had a certain number of shillings. The first had 1/ [one shilling] more than the second, who had 1/ more than the third, and so on. The first gave 1/ to the second, who gave 2/ to the third, and so on, each giving 1/ more than he received, as long as possible. There were then 2 neighbours, one of whom had 4 times as much as the other. How many men were there? And how much had the poorest man at first?

10. A triangular billiard-table has 3 pockets, one in each corner, one of which will hold only one ball, while each of the others will hold two. There are 3 balls on the table, each containing a single coin. The table is tilted up, so that the balls run into one corner, it is not known which. The 'expectation', as to the contents of the pocket, is 2/6. What are the coins?

12. Given the semi-perimeter and the area of a Triangle, and also the volume of the cuboid whose edges are equal to the sides of the Triangle: find the sum of the squares of its sides.

24. If, from the vertices of a triangle *ABC*, the lines *AD*, *BE*, *CF* be drawn, intersecting at *O*: find the ratio *DO/DA* in terms of the two ratios *EO/EB*, *FO/FC*.

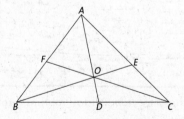

25. If 'ϵ', 'α', 'λ' represent proper fractions; and if, in a certain hospital, 'ϵ' of the patients have lost an eye, 'α' an arm, and 'λ' a leg: what is the least possible number who have lost all three?

[thought out] (57) 22/3/89

To double down part of a given Triangle, making a crease parallel to the base, so that, when the lower corners are folded over it, their vertices may meet.

Let ABC be the given Triangle, & DE the required crease. Let DA'E be the doubled-down piece; & let the lower portions, when folded up, meet at H.

Now $\angle ADE = B$: $\therefore \angle A'DE = B$; $\angle DFB = B$ —

$\therefore \angle DFH = B$ $\therefore \angle HFG = (\pi - 2B)$;

Similarly, $\angle HGF = (\pi - 2C)$;

$\therefore \angle FHG = \pi - (\pi - 2B) - (\pi - 2C)$

$= 2(B + C) - \pi$

$= 2(A + B + C) - 2A - \pi$

$= \pi - 2A$.

$\therefore \triangle HFG$, $\therefore$ sines of $\angle s$ at $H, F, G, = \sin 2A, \sin 2B, \sin 2C$.

Now BF, FG, GC are equal to the sides of this $\triangle$

$\therefore \dfrac{BF}{\sin 2C} = \dfrac{FG}{\sin 2A} = \dfrac{HG}{\sin 2B} = \dfrac{a}{(\sin 2A + \sin 2B + \sin 2C)}$

But $\sin 2A + \sin 2B + \sin 2C = \sin 2A + \sin 2B + \sin 2(\pi - A + B)$

$= \sin 2A + \sin 2B - \sin 2(A + B)$

$= \sin 2A + \sin 2B - \sin(2A + 2B)$

$= \sin 2A (1 - \cos 2B) + \sin 2B (1 - \cos 2A)$

$= \sin 2A . 2 (\sin B)^2 + \sin 2B . 2 (\sin A)^2$

$= 4 \sin A . \sin B . (\sin B . \cos A + \cos B . \sin A)$

$= 4 \sin A . \sin B . \sin$

Whence we can derive BD as is required; & can find D by drawing a $\perp$ from middle of BF. Q.E.F.

32. Sum the series $1·5 + 2·6 +$ &c. (1) to n terms; (2) to 100 terms.

49. If four equilateral Triangles be made the sides of a square Pyramid: find the ratio which its volume has to that of a Tetrahedron made of the Triangles.

57. In a given Triangle describe three Squares, whose bases shall lie along the sides of the Triangle, and whose upper edges shall form a Triangle; (1) geometrically; (2) trigonometrically.

Dodgson's final problem, on 'Transitional Probabilities', has been frequently criticized, but it seems to have been a deliberate joke on his readers:

72. A bag contains 2 counters, as to which nothing is known, except that each is either black or white. Ascertain their colours without taking them out of the bag.

His answer was 'One is black, and the other white.' We present his 'explanation' for those familiar with the study of probability:

We know that, if a bag contained 3 counters, 2 being black and one white, the chance of drawing a black one would be $^2/_3$; and that any other state of things would not give this chance.

Now the chances, that the given bag contains (α) BB, (β) BW, (γ) WW, are, respectively $^1/_4$, $^1/_2$, $^1/_4$.

Add a black counter.

Then the chances that it contains (α) BBB, (β) BWB, (γ) WWB, are, as before, $^1/_4$, $^1/_2$, $^1/_4$.

Hence the chance, of now drawing a black one,

$= ^1/_4·1 + ^1/_2·^2/_3 + ^1/_4·^1/_3 = ^2/_3$.

Hence the bag now contains BBW (since any other state would not give this chance).

Hence, before the black counter was added, it contained BW, i.e. one black counter and one white. Q. E. F.

Sums of Squares

In October 1890, Dodgson was mulling over the problem of finding two squares whose sum is a square (such as $3^2 + 4^2 = 5^2$), and

chanced on a theorem (which seems *true*, though I cannot prove it), that if $x^2 + y^2$ be even, its half is the sum of two squares. A kindred theorem, that $2(x^2 + y^2)$ is always the sum of two squares, also seems true and unprovable.

As an example of the first theorem, $7^2 + 3^2 = 58$ is even, and its half (which is 29) is $5^2 + 2^2$, the sum of two squares. An example of his 'kindred theorem' is that $2 \times (5^2 + 2^2) = 7^2 + 3^2$, which is the sum of two squares.

But these results are certainly not unprovable, as he realized a few days later. They follow immediately from the algebraic formulas

$$\tfrac{1}{2}(x^2 + y^2) = [\tfrac{1}{2}(x + y)]^2 + [\tfrac{1}{2}(x - y)]^2$$

$$2(x^2 + y^2) = (x + y)^2 + (x - y)^2$$

A couple of similar results appear in his *Pillow-Problems*:

14. Prove that 3 times the sum of 3 squares is also the sum of 4 squares.
29. Prove that the sum of 2 different squares, multiplied by the sum of 2 different squares, gives the sum of 2 squares in 2 different ways.

The first of these follows from the general formula

$$3(x^2 + y^2 + z^2) = (x + y + z)^2 + (y - z)^2 + (z - x)^2 + (x - y)^2$$

The answer to the second appears in the Notes at the end of the book.

Number-guessing

Dodgson was keen on number-guessing puzzles. On 3 February 1896, he noted in his diary:

Have been for some days devising an original kind of Number-Guessing Puzzle giving *choice* of numbers to operate with, and have this morning brought it to a very satisfactory form.

An example of this puzzle, which survives in manuscript, takes the form of a dialogue between two people, A and B:

A. "Think of a number."
B. [thinks of 23]

A. "Multiply by 3. Is the result odd or even?"
B. [obtains 69] "It is odd."

A. "Add 5, or 9, whichever you like."

B. [adds 9, & obtains 78]

A. "Divide by 2, & add 1."

B. [obtains 40]

A. "Multiply by 3. Is the result odd or even?"

B. [obtains 120] "It is even."

A. "Subtract 2, or 6, whichever you like."

B. [subtracts 6, and obtains 114]

A. "Divide by 2, & add 29, or 38, or 47, whichever you like."

B. [adds 38, & obtains 95]

A. "Add 19 to the original number, and tack on any figure you like."

B. [tacks on 5, & obtains 425]

A. "Add the previous result."

B. [obtains 520]

A. "Divide by 7, neglecting remainder."

B. [obtains 74]

A. "Again divide by 7. How often does it go?"

B. "Ten times."

A. "The number you thought of was 23."

How did A know?

In order to find B's original number, all A needs to do is to multiply B's final answer by 4 and subtract 15; then, if B's first answer was "Even", subtract 3 more, and if B's second answer was "Even", subtract 2 more.

In fact, Carroll made some slight miscalculations when constructing this puzzle: if B starts with any of the numbers 4, 6 or 8, the answers are "Even", "Even" and "Seven times", so it is impossible for A to know B's original choice. But the matter is easily rectified: in A's seventh instruction, replace 38 and 47 by 33 and 37, and all works out correctly.

Divisibility

An ancient test for deciding whether a given number is divisible by 9 is to add its digits together and see whether the result is divisible by 9. For example, we test whether the number 8 706 528 is divisible by 9 by calculating

$$8 + 7 + 0 + 6 + 5 + 2 + 8 = 36$$

Since 36 is divisible by 9, then so is our original number.

Similarly, we can decide whether a given number is divisible by 11 by alternately adding and subtracting its digits and seeing whether the result is divisible by 11. For example, we test whether the above number is divisible by 11 by calculating

$$8 - 7 + 0 - 6 + 5 - 2 + 8 = 6$$

Since 6 is not divisible by 11, then nor is our original number — indeed, we obtain 6 as a remainder when we carry out the division.

In the autumn of 1897, Dodgson made the following entries in his diary:

> *27 September*: *Dies notandus*. Discovered rule for dividing a number by 9, by mere addition and subtraction. I felt sure there must be an analogous one for 11, and found it, and proved first rule by Algebra, after working about nine hours!
>
> *28 September*: *Dies cretâ notandus!* I have actually *superseded* the rules discovered yesterday! My new rules require to ascertain the 9-remainder, and the 11-remainder, which the others did *not* require: but the new ones are much the quickest. I shall send them to the *Educational Times*, with date of discovery.

Dodgson's methods for ascertaining divisibility by 9 and 11 were more sophisticated than those given above, since they yielded not only the remainder but also the quotient. For example, on using his methods to divide 8 706 528 by 11, he would quickly find both the remainder 6 and the quotient, 791 502.

He then became more ambitious, extending his methods to the division of any number by 13, 17, 19, 41, and a whole host of other numbers, including all those within 10 of a power of 10, such as 107 and 991. Within a couple of months he was able to write:

26 November: Completed my rules for dividing by divisors of the form $(ht^n \pm 1)$, where h is > 1.

30 November: 4 a.m. have just completed *another* new rule — for divisors of the form $(ht^n \pm k)$: i.e. I can find the *Remainder*, but not yet the *Quotient*.

7 December: I have since completed the above rule — by finding the *Quotient*.

He illustrated his rule by speedily dividing 5 984 407 103 826 by 6993 (= $7 \times 10^3 - 7$), yielding the quotient as 855 281 849 and the remainder as 6373.

Right-angled Triangles

Dodgson's last piece of mathematics was a problem that involved finding three right-angled triangles of the same area. His diary entry for 19 December 1897 reads:

> Sat up last night till 4 a.m., over a tempting problem, sent me from New York, "to find three equal rational-sided right-angled triangles". I found *two*, whose sides are 20, 21, 29; 12, 35, 37: but could not find *three*.

'Rational-sided' means that all the sides are whole numbers or fractions, while 'equal' means that they have equal areas. Dodgson's triangles are shown below: each has area 210.

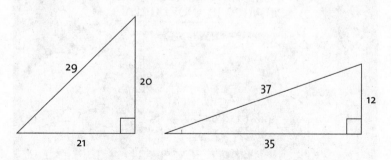

In fact, he was closer than he knew. The smallest solution of this problem consists of the three right-angled triangles with sides

40, 42, 58,   24, 70, 74   and   15, 112, 113

with common area 840; the first two of these have sides twice the length of those that Dodgson found. It is now known that there are infinitely many solutions to this problem; another is:

$$105, 208, 233, \quad 91, 120, 218 \quad \text{and} \quad 56, 390, 394$$

Epilogue

Just four days later, Dodgson was off to Guildford to spend Christmas at The Chestnuts with his sisters. His brief diary entry for 23 December 1897 was the last that he wrote:

I start for Guildford by the 2.07 today.

While in Guildford he worked hard on Part II of his *Symbolic Logic*, hoping to finish it. On 6 January he developed a feverish cold which rapidly developed into severe bronchial influenza. He died on 14 January 1898 at the age of sixty-five, and following a simple funeral was buried in The Mount Cemetery in Guildford. In the same week, Henry Liddell, whose daughter Alice had been so much a part of Dodgson's life, also died, and a joint memorial

Stained-glass window in Daresbury Church, Cheshire

service was held for them at Christ Church Cathedral on Sunday 23 January. As Dean Paget recalled:

> Within the last ten days Christ Church has lost much. And though the work that bore the fame of Lewis Carroll far and wide stands in distant contrast with the Dean's, still it has no rival in its own wonderful and happy sphere; and in a world where many of us laugh too seldom, and many of us laugh amiss, we all owe much to one whose brilliant and incalculable humour found us fresh springs of clear and wholesome and unfailing laughter.

So let us leave the final word with Dodgson's alter ego, Lewis Carroll. On 25 June 1857, while sitting alone in his room listening to the music from a Christ Church ball, he composed a double acrostic puzzle, one of whose verses has sometimes been quoted as his own mathematical self-portrait:

> *Yet what are all such gaieties to me*
> *Whose thoughts are full of indices and surds?*
> $$x^2 + 7x + 53 = 11/3$$

Notes and References

Further Reading

The abbreviated titles in square brackets are used in the Notes that follow.

Biographical accounts of Charles Dodgson's life and works appear in a number of sources. Two useful books were written shortly after he died by his nephew:

Stuart Dodgson Collingwood, *The Life and Letters of Lewis Carroll*, Fisher Unwin, 1898. [*Life*]

Stuart Dodgson Collingwood, *The Lewis Carroll Picture Book*, Fisher Unwin, 1899. [*Picture Book*]

Of the many biographies of Dodgson, one of the best known is:

Morton N. Cohen, *Lewis Carroll: A Biography*, Macmillan, 1995. [*Biography*]

Much useful information about his life can be found in his diaries and letters:

Lewis Carroll's Diaries: The Private Journals of Charles Lutwidge Dodgson, in ten volumes, edited by Edward Wakeling, The Lewis Carroll Society, 1993–2007. [*Diaries*]

The Selected Letters of Lewis Carroll, in two volumes, edited by Morton N. Cohen with the assistance of Roger Lancelyn Green, Macmillan, 1979. [*Letters 1 and 2*]

There are several collections of his writings. One of the best known is:

The Complete Works of Lewis Carroll, Penguin Books, 1988. [*Works*]

We use the following abbreviations of his works for children:

AAW	*Alice's Adventures in Wonderland*
HS	*The Hunting of the Snark*
SB	*Sylvie and Bruno*
SBC	*Sylvie and Bruno Concluded*
TLG	*Through the Looking-Glass*

Several of his works were reissued as paperback books by Dover Publications, New York; the first two appeared in 1958 under the heading of *The Mathematical Recreations of Lewis Carroll*, and the third appeared in 1973:

Pillow-Problems and *A Tangled Tale*.

Symbolic Logic and *The Game of Logic*.

Euclid and his Modern Rivals, with a new Introduction by H.S.M. Coxeter.

Much of his Oxford output — both mathematical and of a more general nature — appears in the three volumes of *The Pamphlets of Lewis Carroll*, published for the Lewis Carroll Society of North America, and distributed by the University of Virginia:

The Oxford Pamphlets, Leaflets, and Circulars of Charles Lutwidge Dodgson, compiled, with notes and annotations, by Edward Wakeling, 1993. [*Pamphlets 1*]

The Mathematical Pamphlets of Charles Lutwidge Dodgson and Related Pieces, compiled, with introductory essays, notes, and annotations, by Francine F. Abeles, 1994. [*Pamphlets 2*]

The Political Pamphlets and Letters of Charles Lutwidge Dodgson and Related Pieces: A Mathematical Approach; compiled, with introductory essays, notes, and annotations, by Francine F. Abeles, 2001. [*Pamphlets 3*]

Other general sources are listed below at the beginning of the appropriate Fits.

Notes

In the many quoted extracts from his writings in this book we have followed Dodgson's sometimes rather idiosyncratic style as closely as possible.

Frontispiece
The picture and text, from a letter to Margaret Cunnynghame, appear in *Life*, pp. 424–425, and *Letters 1* (30 January 1868), pp. 112–113; in the original letter the verse appears as continuous text.

Introduction: From Gryphons to Gravity
Full versions of the extracts adapted here appear in *Works*.
Opening quotation: *AAW*, Ch. XII, 'Alice's Evidence' (*Works*, p. 114).

Scene 1: The Mock Turtle's Education
AAW, Ch. IX, 'The Mock Turtle's Story' (*Works*, pp. 93–95).

Scene 2: Humpty Dumpty's Cravat
TLG, Ch. VI, 'Humpty Dumpty' (*Works*, pp. 192–196).

Scene 3: Alice's Examination
TLG, Ch. IX, 'Queen Alice' (*Works*, pp. 230–233).

Scene 4: What's in a Name?
AAW, Ch. VI, 'Pig and Pepper'; *TLG*, Ch. V, 'Wool and Water', Ch. VII, 'The Lion and the Unicorn', and Ch. VIII, 'It's My own Invention' (*Works*, pp. 66–67, 184, 204–205, 207, 224).

Scene 5: The Beaver's Lesson

HS, Fit the Fifth, 'The Beaver's Lesson' (*Works*, pp. 691–692).

Scene 6: Map-making

HS, Fit the Second, 'The Bellman's Speech'; *SBC*, Ch. XI, 'The Man in the Moon' (*Works*, pp. 556–557, 683).

Scene 7: Fortunatus's Purse

SBC, Ch. VII, 'Mein Herr' (*Works*, pp. 521–523). This scene is based on an earlier dramatization by the late John Fauvel.

Scene 8: A Question of Gravity

SBC, Ch. VII, 'Mein Herr' (*Works*, pp. 524–525); *AAW*, Ch. 1, 'Down the Rabbit-Hole'; *SB*, Ch. VIII, 'A Ride on a Lion' (*Works*, pp. 17, 311–313).

Fit the First: The Children of the North

Introductions to Dodgson's early years can be found in *Life*, Ch. I, *Biography*, Ch. 1, and *Diaries*, Vol. 1. There is also a good account in Anne Clark Amor, *Lewis Carroll: Child of the North*, The Lewis Carroll Society (on behalf of St Peter's Church, Croft), 1995.

Daresbury

The quotations and the poem appear in *Life*, pp. 7, 8, 11–13, 15. A slightly different version of the poem, entitled 'Faces in the Fire', is in *Works*, p. 875:

> An island-farm — broad seas of corn,
> Stirred by the wandering breath of morn —
> The happy spot where I was born.

Croft

The poems, entitled 'Lays of Sorrow No. 2' and 'Facts', appear in *Works*, pp. 703, 712. The punctuation in 'Facts' is Charles's own.

The quotations are from *Life*, pp. 19, 21, 24.

Richmond

Tate's two comments are in *Life*, pp. 25–26.

The page of geometry appeared in *The Colophon*, New Graphic Series No. 2, New York, 1939; it is one of the earliest known examples of Charles Dodgson's handwriting. Our selection of problems is from Dodgson's copy of the 1842 edition of Walkingame's book. The Latin inscription is reproduced in Derek Hudson, *Lewis Carroll*, Constable, 1954, p. 45.

Rugby

Charles's reminiscences appear in *Life*, pp. 30–31.

Information about his prizes comes from *Diaries 1*, pp. 23–24.

Charles's letters to Elizabeth Dodgson are in *Letters 1* (9 October 1848 and 24 May 1849), pp. 7, 10; and *Life*, p. 28.

The quotations from Mayor and Tait appear in *Life*, p. 29.

Marking time

'Difficulty No. 1' is reproduced in full in *Picture Book*, pp. 4–5; and *Works*, pp. 1115–1116.

Dodgson's contribution to *The Illustrated London News* on 18 April 1857 is to be found in John Fisher (ed.), *The Magic of Lewis Carroll*, Nelson, 1973, pp. 23–24.

The 'pleasant party' quotation appears in *Life*, p. 85.

The Alice quotations are adapted from *AAW*, Ch. VI, 'Pig and Pepper', and Ch. VII, 'A Mad Tea-party' (*Works*, pp. 61–62, 71).

The letter to Elizabeth Dodgson is from *Diaries 1* (24 October 1849), p. 11.

'Difficulty No. 2' appears in *Picture Book*, p. 6; and *Works*, pp. 1108–1109.

Fit the Second: Uppe toe mine Eyes yn Worke

Much of the material in this Fit is based on *Life*, Ch. II; and *Biography*, Ch. 2.

Introduction

Dodgson's remarks on Christ Church appear in 'Isa's Visit to Oxford, 1888', in *Picture Book*, p. 323.

An Oxford Undergraduate

Dodgson's verse, from 'A Bacchanalian Ode', concludes *The Vision of the Three T's* (*Works*, p. 1053; and *Pamphlets 1*, p. 100).

Dr Jelf's letter is in *Life*, p. 46.

For information about Oxford University and its examinations, see the annual editions of the *Oxford University Calendar* for 1850–1855.

A Trio of Examinations

Responsions

The 'long table' quote is from *The Christ Church Commoner* (1851), Chapter 1, reprinted in the Lewis Carroll Circular, No. 1 (1973).

Dodgson's 'Olde Englishe' letter to Louisa appears in *Letters 1* (10 June 1851), p. 15.

The quotation is from *Life*, p. 51; and *Letters 1* (to Elizabeth Dodgson, 5 July 1851), p. 17.

Moderations

The Revd Baden Powell was the father of the founder of the Scout movement.

The Dodgson quotations are from *Letters 1* (to Elizabeth Dodgson, 24 June and 9 December 1852), pp. 19–20, 22.

The Pusey and Revd Dodgson quotations appear in a letter from the latter (*Life*, pp. 53, 55).

Dodgson's letter to his cousin W.M. Wilcox appears in *Letters 1* (10 September 1885), p. 602.

Finals

The Hatter's song is from *AAW*, Ch. VII, 'A Mad Tea-party' (*Works*, p. 72).

The quotations appear in *Letters 1* (to Mary Dodgson, 23 August 1854), p. 29; and in the *St James's Gazette*, 11 March 1898 (reminiscence of Thomas Fowler); *Life*, p. 221; and *Diaries 7* (24 November 1882), p. 498.

The hints for studying are from a letter to Edith Rix and appear in *Life*, pp. 240–241.

Dodgson's letter to Mary appears in *Life*, p. 58; and *Letters 1* (13 December 1854), pp. 29–30.

Fit the Third: Successes and Failures

Many of the extracts in this Fit are in the *Diaries*; and in *Life*, Ch. II, and *Biography*, Ch. 3.

Dodgson's photographs have been published in two main collections:

Morton N. Cohen, *Reflections in a Looking Glass*, Aperture, 1998.

Roger Taylor and Edward Wakeling, *Lewis Carroll Photographer*, Princeton University Press, 2002.

Introduction

The letter to Mary appears in *Letters 1* (13 December 1854), p. 30.

The Senior Scholarship

The diary entries are from *Diaries 1* (15 January, 17 February and 22–24 March 1855), pp. 56, 62, 77–78.

College Teaching

The diary entries are from *Diaries 1* (30 January, 5 March, 26 April and 14 May 1855), pp. 60, 70, 87, 97.

The letter to Henrietta and Edwin appears in *Letters 1* (31 January 1855), p. 31; and *Picture Book*, pp. 198–199. In the original letter the 'lecture' appears as continuous text.

The 'proposition' is from *The Dynamics of a Parti-cle*, 1865 (*Works*, p. 1023; *Picture Book*, p. 71; and *Pamphlets 1*, p. 33).

A New Appointment

The verse is an Oxford variant on the original by Cecil Spring-Rice in *The Balliol Rhymes*, Basil Blackwell, Oxford, 1939, which begins:

> I'm the Dean of Christ Church; — Sir
> There's my wife, look well at her.

The diary extracts are from *Diaries 1* (22 June 20 August and 31 December 1855), pp. 79, 105, 136; and *Diaries 7* (14 July and 30 November 1881), pp. 349, 381. The 'Hints for Etiquette' appear in *Picture Book*, pp. 33–34. The letter from Dodgson's father (21 August 1855) is reproduced in *Life*, p. 61.

A Spot of Schoolteaching

The diary entries are from *Diaries 1* (8, 10, 16 July 1855), pp. 108–111.

The addition sum

Note that 5314 = 9999 − 4685 and 2937 = 9999 − 7062; in general, I subtract from 9999 whatever numbers you choose. The final total is then the original number (2879) plus (9999 + 9999) = 22877, as predicted.

Counting alternately up to 100

Note that 7 = 11 − 4 and 3 = 11 − 8; in general, I subtract from 11 whatever number you choose and add the result to the total. Thus, the numbers I announce increase by 11 each time: 1, 12, 23, 34, 45, 56, 67, 78, 89 and 100. Dodgson later developed this idea into a game called *Arithmetical Croquet*.

The '9' trick

If you take any number and subtract its reverse, the result can always be divided exactly by 9, and so can the sum of its digits. When you remove one digit and tell me the sum of the remaining digits, I then calculate how much to add to increase this sum to a multiple of 9; for example, if you give me the sum 13, I must add 5 to increase the total to 18 (= 2 9), so the number you removed was 5.

Dodgson as a Teacher

The quotations appear in Morton N. Cohen, *Lewis Carroll: Interviews and Recollections* (letters to *The Times* from John H. Pearson, Herbert Maxwell and Watkin H. Williams, 19, 22 December 1931 and 12 January 1932), Macmillan, 1989, pp. 75–77. The *pons asinorum* is described in Fit the Fourth. The diary entries are from *Diaries 2* (12, 26 November 1856), pp. 113–115, 119.

Poems and Photographs

The diary entry is from *Diaries 2* (11 February 1856), p. 39; the pen-name Lewis Carroll was chosen on 1 March.

This extract from *The Song of Hiawatha* introduces Part XXII, 'Hiawatha's Departure'.

'Hiawatha's Photographing' exists in several different versions; that used here is the original from *The Train*, Vol. IV, July–December 1857; for a later version, see *Works*, pp. 768–772.

The quotations appear in Morton N. Cohen, *Lewis Carroll: Interviews and Recollections* (from Ella Bickersteth), Macmillan, 1989, p. 190; and *Picture Book*, p. 224; and from Helmut Gernsheim, *Lewis Carroll — Photographer*, Dover Publications, 1969, p. 28.

Ciphers

The diary entries are from *Diaries 2* (15 February 1856), p. 42; and *Diaries 3* (26 February 1858), p. 161.

The two decoded messages are 'Beware the Jabberwock, my Son!' and ''Twas brillig, and the slithy toves', from *TLG*, Ch. 1, 'Looking-Glass House' (*Works*, pp. 140–141).

Fit the Fourth: . . . in the Second Book of Euclid

Several of the extracts in this chapter appear in *Pamphlets 1, 2*.

Euclid and his Modern Rivals was published by Macmillan in March 1879; a *Supplement* and second edition were published in 1885.

Introduction

The letter to Elizabeth appears in *Letters 1* (9 October 1848), p. 7.

The parallelepiped verse is in 'Sad Souvenaunce', Canto VII of *Phantasmagoria* (*Works*, p. 765).

The geometrical word-play is from *SB*, Ch. XII, 'A Musical Gardener' (*Works*, p. 338).

The W.S. Gilbert quotation is from *The Pirates of Penzance*, 1879.

The Pythagoras quotation is from Dodgson's *Curiosa Mathematica*, Part I: *A New Theory of Parallels*, 1888, p. x.

Here's Looking at Euclid

The diary entry is from *Diaries 1* (16 April 1855), pp. 82–83.

Robert Potts's edition of Euclid's *Elements* was first published in 1847; Dodgson's copy was a later edition, dated 1860.

Dodgson's descriptions of a postulate and an axiom are from *Notes on the First Two Books of Euclid* (1860) (*Pamphlets 2*, p. 36). The given definitions, postulates and axiom appear at the beginning of Euclid's *Elements*, Book I.

Dodgson's Pamphlets
The quotation is from the preface of *Euclid, Books I, II*.
The Potts quotation is from a letter from Robert Potts dated 20 March 1861.

The Dynamics of a Parti-cle
The quotations from *The Dynamics of a Parti-cle* appear in *Works*, pp. 1018–1019, 1024–1025; *Picture Book*, pp. 63–64, 72–74; and *Pamphlets 1*, pp. 28–29, 35–36.

The Euclid Debate
The Whewell quotation is from his *Of a Liberal Education*, Parker, 1845, p. 30.
The story about the Oxford examination student appears in J. Pycroft, *Oxford Memories*, Bentley, 1886, p. 82.
Sylvester's Presidential Address to the Mathematics and Physics Section of the British Association was reprinted in *Nature* 1 (1869–1870), 261; the quotation 'deeper than e'er plummet sounded' is from Shakespeare's *The Tempest* (V i 56). The De Morgan quotation appears in his review of J.M. Wilson's *Elementary Geometry* in *The Athenaeum* 2125 (18 July 1868), 71–73.

Euclid and his Modern Rivals
The extracts from *Euclid and his Modern Rivals* appear in the preface, pp. ix–x, and on pp. 1–2, 11, 182–183, 211–212, 225. Quotations within them are from Shakespeare: 'I think we do know the sweet Roman hand' is from *Twelfth Night* (III iv 31), 'Now infidel, I have thee on the hip' from *The Merchant of Venice* (IV i 335), and 'Our revels now are ended . . .' from *The Tempest* (IV i 148).
The rival texts are, in the order in which Dodgson deals with them:
    A.-M. Legendre, *Éléments de Géométrie*, 16th edition (1860).
    W.D. Cooley, *The Elements of Geometry, Simplified and Explained* (1860).
    Francis Cuthbertson, *Euclidian* [sic] *Geometry* (1874).
    Olaus Henrici, *Elementary Geometry: Congruent Figures* (1879).
    J.M. Wilson, *Elementary Geometry*, 2nd edition (1869).
    Benjamin Peirce, *An Elementary Treatise on Plane and Solid Geometry* (1872).
    W.A. Willock, *The Elementary Geometry of the Right Line and Circle* (1875).
    W. Chauvenet, *A Treatise on Elementary Geometry* (1876).
    Elias Loomis, *Elements of Geometry*, revised edition (1876).
    J.R. Morell, *Euclid Simplified* (1875).
    E.M. Reynolds, *Modern Methods in Elementary Geometry* (1868).
    R.F. Wright, *The Elements of Plane Geometry*, 2nd edition (1871).
    Syllabus of the Association for the Improvement of Geometrical Teaching (1878).

The quotation from *Euclid and his Modern Rivals* appears on p. 42.

'The cock doth craw, the day doth daw' is a line from a traditional Scottish ballad, 'The Wife of Usher's Well'. The line is spoken by the ghosts of the Wife's two sons, who tell her that at daybreak they must depart.

Dodgson's Hexagon

The quotation from *The Dynamics of a Parti-cle* is to be found in *Works*, pp. 1016–1017; *Picture Book*, p. 60; and *Pamphlets 1*, p. 27.

The 'Puck' quotation is from *Curiosa Mathematica*, Part I: *A New Theory of Parallels*, 1888, p. xix.

Squaring the Circle

The opening quotation is from Augustus De Morgan, *A Budget of Paradoxes*, Dover, 1954, p. 151 (1st edn, Longmans, Green & Co., 1872).

Dodgson's circle-squaring quotations are in *Curiosa Mathematica*, Part I: *A New Theory of Parallels* (1888), p. xi; and *Pamphlets 2*, pp. 144–145. The moonshine quotation appears in *Life*, p. 217.

Fit the Fifth: Send Me the Next Book . . .

For this period of Dodgson's career, see *Life*, Ch. III. Several extracts in this Fit appear in *Pamphlets 1, 2*.

Alice's Adventures in Wonderland was published by Macmillan in July 1865, but the illustrator John Tenniel was dissatisfied with the printing of the pictures and the edition was recalled; a revised printing appeared in November 1865. *An Elementary Treatise on Determinants* was published by Macmillan on 10 December 1867.

Introduction

Dodgson denies the Queen Victoria story on the title page of the second edition of his *Symbolic Logic* (1896).

Dodgson the Deacon

Dodgson's letter to the Diocesan Registrar appears in *Letters 1* (5 August 1861), p. 50.

Dodgson's letter to his cousin and godson W.M. Wilcox is reproduced in *Letters 1* (10 September 1885), pp. 602–603.

Bishop Wilberforce's views on the theatre are from *Life*, p. 74.

Dodgson's reading difficulties are described in *Letters 2* (to H.L.M. Walters, 5 January 1898), p. 1154.

Collingwood's comments on Dodgson's sermons appear in *Life*, pp. 76, 78.

More Pamphlets

The letter to Mary appears in *Letters 1* (20 February 1861), p. 48.

The extract from *Circular to Mathematical Friends* is in *Pamphlets 2*, p. 351.

Letters to Child-friends

Dodgson's letter-writing comment is from *Diaries 8* (20 March 1884), p. 95.

The inkstand letter appears in *Letters 1* (to Marion Richards, 26 October 1881), p. 440.

The letter to Isa Bowman is to be found in *Letters 2* (14 April 1890), pp. 785–786.

The reminiscence of Oxford (by Winifred Stevens) appears in *Picture Book*, p. 200.

Dodgson's remark that he was 'fond of children (except boys)' appears in *Letters 1* (to Kathleen Eschwege, 24 October 1879), p. 351.

The letter to Wilton Rix is in *Letters 1* (20 May 1885), p. 577. The argument uses the fact that $x^2 - y^2 = (x + y)(x - y)$; since one must not divide by zero, the error occurs when he divides each side of the equation by $(x - y)$, which is 0 since x and y are both 1.

The Duckworth quotation is from *Picture Book*, pp. 358, 360.

The Alice quotation appears in *AAW*, Ch. 2, 'The Pool of Tears' (*Works*, p. 30).

Dodgson's *Determinants*

The diary entries are from *Diaries 5* (27–28 February and 15, 29 March 1866), pp. 132–133, 206–207.

Dodgson's paper had the title 'Condensation of determinants, being a new and brief method for computing their arithmetical values' and appeared in the *Proceedings of the Royal Society* 15 (1866), pp. 150–155. For an explanation of his method, see Adrian Rice and Eve Torrance, '"Shutting up like a telescope": Lewis Carroll's "curious" condensation method for evaluating determinants', *College Mathematical Journal* 38 (February 2007), 85–95.

The diary entry appears in *Diaries 5* (11 July 1867), p. 253.

Dodgson's Continental trip is described in *Life*, pp. 111–126; the comment about Dr Liddon's fame as a preacher is on p. 121. Dodgson's own record of the tour appears in *Diaries 5* (12 July–13 September 1867), pp. 255–369. The tale of the parrot is from the entry for 22 July 1867, on pp. 277–278.

University Whimsy

The diary entry appears in *Diaries 5* (3 March 1865), p. 54.

The extracts from *The New Method of Evaluation as Applied to* π are in *Works*, pp. 1011–1016; *Picture Book*, pp. 45–57; and *Pamphlets 1*, pp. 19–25.

The satirical document was a pamphlet entitled *The Blank Cheque, A Fable*, 1874, protesting the University's plans to build the New Examination Schools without properly estimating the expense; the extract appears in *Works*, p. 1057; *Picture Book*, p. 146; and *Pamphlets 1*, p. 114.

'The Offer of the Clarendon Trustees', dated 6 February 1868, is in *Works*, pp. 1009–1011.

Fit the Sixth: Meat-safes, Majorities and Memory

For this period of Dodgson's career, see *Life*, Chs. IV and V.

Much information about Dodgson's work on majority voting, lawn tennis tournaments and parliamentary representation can be found in *Pamphlets 3*, and in Francine Abeles, 'The mathematical-political papers of C. L. Dodgson', in *Lewis Carroll: A Celebration* (ed. Edward Guiliano), Clarkson N. Potter, 1982, pp. 195–210 (referred to below as *Abeles*).

Introduction

The Alice quotation is from *AAW*, Ch. VII, 'A Mad Tea-party' (*Works*, p. 75). Dodgson's recollection appears in *Life*, p. 131.

College Life

Dodgson's verse, from 'A Bacchanalian Ode', concludes *The Vision of the Three T's* (*Works*, p. 1053, and *Pamphlets 1*, p. 100).

The quotations are from 'The New Belfry of Christ Church, Oxford' (*Works*, pp. 1026–1036, and *Pamphlets 1*, pp. 69–79); Ariel's song is from Shakespeare's *The Tempest* (I ii 394).

Voting in Elections

The Marquis de Condorcet's *Essai sur l'application de l'analyse à la probabilité des décisions rendues à la pluralité des voix* dates from 1785.

The diary entries are from *Diaries 6* (13, 18 December 1873), pp. 306, 307.

Dodgson's *A Discussion of the Various Methods of Procedure in Conducting Elections* appears in *Pamphlets 3*, pp. 33–41.

Dodgson's *A Method of Taking Votes on More than Two Issues* appears in *Pamphlets 3*, pp. 47–58.

Memoria Technica

The *Memoria Technica* appears in *Life*, pp. 268–269; and Martin Gardner, *The Universe in a Handkerchief*, Springer-Verlag, 1996, pp. 33–35.

Evelyn Hatch's examples are from her book *A Selection from the Letters of Lewis Carroll to his Child-Friends*, Macmillan, 1933, pp. 233, 234.

The diary entries appear in *Diaries 6* (27/28 October 1875), p. 428 and the notes on pp. 429–431.

Endings and Beginnings

The diary entries are from *Life*, pp. 218–219; and *Diaries 7* (1 February and 6 March 1880; 14 July, 18, 21 October and 30 November 1881; and 8 December 1882), pp. 239, 248, 349, 371, 373, 380, 500–501; and *Diaries 8* (29 March 1885), pp. 179–183.

Lawn Tennis Tournaments

The quotations from *Lawn Tennis Tournaments* appear in *Works*, p. 1082; and *Pamphlets 3*, p. 72.

Parliamentary Representation

The quotation, from *Purity of Election* (1881), appears in *Pamphlets 3*, p. 137. The table is from *Abeles*, p. 201.

The extracts from *The Principles of Parliamentary Representation* are adapted from *Pamphlets 3*, pp. 178–179, 198–199.

The remark by Michael Dummett is from *Voting Procedures*, Oxford, 1984, p. 5, and appears in *Pamphlets 3*, p. 31.

Fit the Seventh: Puzzles, Problems and Paradoxes

A Tangled Tale (referred to below as *TT*) was published by Macmillan in 1885 and republished in *The Mathematical Recreations of Lewis Carroll: Pillow-Problems and A Tangled Tale*, Dover Publications, 1958.

Dodgson's puzzles and paradoxes are discussed at length in the following publications:

> *Lewis Carroll's Games and Puzzles* and *Rediscovered Lewis Carroll Puzzles*, newly compiled and edited by Edward Wakeling, Dover Publications (in conjunction with the Lewis Carroll Birthplace Trust, Daresbury, Cheshire), 1992, 1995 (referred to below as *Puzzles 1, 2*).
>
> John Fisher (ed.), *The Magic of Lewis Carroll*, Nelson, 1973.
>
> Martin Gardner, *The Universe in a Handkerchief*, Springer-Verlag, 1996 (referred to below as *Universe*).

Introduction

The origin of *The Hunting of the Snark* is recounted in 'Alice on the stage', *The Theatre*, April 1887, and is reprinted in Martin Gardner's *The Annotated Snark*, Penguin, 1962, p. 16.

The last stanza of *The Hunting of the Snark*, from Fit the Eighth, is in *Works*, p. 699.

The letter to Birdie appears in *Letters 1* (6 April 1876), p. 246; and *Picture Book*, pp. 209–210.

A Tangled Tale

The Alice quotation is from *AAW*, Ch. III, 'A Caucus Race and a Long Tale' (*Works*, p. 36).

The poem 'To My Pupil' and the author's purpose in writing *TT* are to be found in *Works*, pp. 881–882.

'Knot I: Excelsior' appears in *TT*, pp. 1–3 (*Works*, pp. 883–884).

The solutions and discussion for Knot I, the 'class list' and the reply to SCRUTATOR's complaint are in *TT*, pp. 77–78, 81, 89 (*Works*, pp. 922, 925, 930–931).

The statements of the problems in Knots II–X appear in *TT*, pp. 84, 86, 90, 96, 102, 106–107, 112, 132, 133, 135–137, 142, 146 (*Works*, pp. 926–928, 931, 934, 938, 940–941, 944, 957, 958–960, 963, 965).

Carroll's answers to Knots II–X are as follows (full explanations appear in *TT* and *Works*):

Knot II

Problem 1: One. In this genealogy, males are denoted by capitals, and females by small letters. The Governor is E and his guest is C.

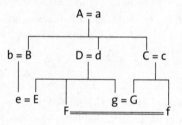

Problem 2: From No. 9.

Knot III

(1) 19.   (2) The easterly traveller met 12; the other 8.

Knot IV

5½, 6½, 7, 4½, 3½.

Knot V

10 pictures; 29 marks; arranged thus:

×	×	×	×	×	×	×	×	×	○
×	×	×	×	×	○	○	○	○	○
×	×	○	○	○	○	○	○	○	○

Knot VI

Problem 1: They went that day to the Bank of England. *A* stood in front of it, while *B* went round and stood behind it.

Problem 2: The order is *M, L, Z*.

Knot VII

(1) 8d.; (2) 1s. 7d.

Knot VIII

Problem 1: Place 8 pigs in the first sty, 10 in the second, nothing in the third, and 6 in the fourth: 10 is nearer ten than 8; nothing is nearer ten than 10; 6 is nearer ten than nothing; and 8 is nearer ten than 6.

Problem 2: In 6¼ minutes.

Knot IX

Problem 1: Lardner means by "displaces," "occupies a space which might be filled with water without any change in the surroundings." If the portion of the floating bucket, which is above the water, could be annihilated, and the rest of it transformed into water, the surrounding water would not change its position: which agrees with Lardner's statement. [Dionysius Lardner was a British physicist and astronomer who lived from 1793 to 1859.]

Problem 2: No: this series can never reach 4 inches, since, however many terms we take, we are always short of 4 inches by an amount equal to the last term taken. [This problem is a variation on Zeno's paradox; see Fit the Eighth.]

Problem 3: 60, 60½.

Knot X

Problem 1: Ten.

Problem 2: 15 and 18.

Carroll's Puzzles

Arithmetical Puzzles

The 'magic number' curiosity appears in *Picture Book*, p. 269; and *Puzzles 1*, p. 15; it may be the 'number repeating puzzle' that Carroll showed at Guildford High School: see *Diaries 9* (26 January 1897), p. 293.

The description of the children's party appears in *Biography*, p. 208.

The 1089 puzzle has given its name to a popular mathematics book, David Acheson's *1089 and All That*, Oxford, 2001. The money version appears in *Picture Book*, p. 269; and *Puzzles 2*, p. 41.

If the original number (or sum of money) is *abc*, (or £*a* b*s*. c*d*.), where *a* is larger than *c*, then the calculations are as follows:

Hundreds	Tens	Units		£	s	d
a	b	c		a	b	c
$-\quad c$	b	a	$-$	c	b	a
$a-c-1$	9	$10+c-a$		$a-c-1$	19	$12+c-a$
$+\ 10+c-a$	9	$a-c-1$	$+$	$12+c-a$	19	$a-c-1$
10	8	9		£12	18s.	11d.

Money problem This appears in *Picture Book*, pp. 317–318; and *Puzzles* 2, p. 49. Pooling the money gives 10s., 5s., 4s., 2s. 6d., 2s., 1s., 6d., 4d., 3d. and 1d. The customer takes 5s. 3d. (5s. and 3d.); the friend who loaned 7s. 1d. removes this amount (4s., 2s. 6d., 6d. and 1d.); the shopkeeper, who started with 6s. 1d. and receives 7s. 3d., now has 13s. 4d. (10s., 2s., 1s. and 4d.).

Brandy and water This problem, as recalled by Viscount Simon, appears in Morton N. Cohen, *Lewis Carroll: Interviews and Recollections*, Macmillan, 1989, p. 67; and *Puzzles* 2, p. 52. We can calculate the amount of each liquid transferred: 50/51 of the brandy from the first tumbler to the second and 50/51 of water from the second tumbler to the first. But, as Carroll observed, it is much easier to note that each tumbler ends up with 50 spoonfuls of liquid, and so the amount of brandy in the second tumbler must equal the amount of water in the first.

Geometrical Puzzles

The square window This problem was posed in a letter from Carroll to Helen Feilden and appears in *Letters* 1 (15 March 1873), p. 187; *Picture Book*, p. 214; and *Puzzles* 1, p. 36.

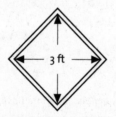

The extra square This problem is in *Picture Book*, pp. 316–317; and *Puzzles* 2, p. 7. If you draw the second diagram carefully you will notice that there is a very thin gap in the middle, whose area equals that of one square.

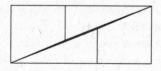

The three squares problem Carroll's letter to Isabel Standen appears in *Letters 1* (22 August 1869), p. 138; for her recollection of the problem see *Life*, p. 370. Here is one solution:

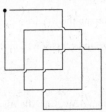

Painting cubes This problem, due originally to Major Percy A. MacMahon, was found among 'Bat' Price's papers after Dodgson's death, and appears in *Puzzles 2*, p. 18. The solution is as follows. Choose a face and paint it; then the opposite face can be coloured with any of the remaining five colours. Now paint another face; then there are just six ways of painting the remaining three faces. This gives a total of 5 × 6 = 30 different ways of painting the faces.

Clock face This problem was also found among Price's papers. The answer is $27^9/_{13}$ minutes past six o'clock, and a full solution is reproduced in *Puzzles 1*, pp. 66–67.

River-crossing Puzzles
The earliest known river-crossing puzzle, due to the ecclesiastic and scholar Alcuin of York in the ninth century, involved a wolf, a goat and a cabbage.

The letter to Jessie Sinclair appears in *Letters 1* (22 January 1878), p. 300; and *Picture Book*, p. 205. The 'awful story' is in *Letters 1* (to Helen Feilden, 15 March 1873), p. 187; and *Picture Book*, pp. 214–215.

For Edward Craig's reminiscence from his *Index to the Story of my Days*, Hulton Press, 1957, p. 32, see Morton N. Cohen, *Lewis Carroll: Interviews and Recollections*, Macmillan, 1989, p. 153.

The problem of the captive Queen appears in *Picture Book*, p. 318. Here is a solution:

1. Send down the weight — the empty basket comes up.
2. Send down the son — bring up the weight.
3. Send down the weight.
4. Send down the daughter — bring up the son and the weight.
5. Send down the weight.
6. Send down the son — bring up the weight.
7. Send down the Queen and the weight — bring up the daughter and the son.

8. Send down the son — bring up the weight.
9. Send down the weight.
10. Send down the daughter — bring up the son and the weight.
11. Send down the weight.
12. Send down the son — bring up the weight.

The variation with the pig, dog and cat is solved in *Puzzles* 2, pp. 65–66.

Other Recreations
The Number 42
The Court scene quotation is from *AAW*, Ch. XII, 'Alice's Evidence' (*Works*, p. 113).
The *Snark* quotations appear in *HS*, in the Preface and Fit the First, 'The Landing' (*Works*, pp. 677–678, 681).
The *Phantasmagoria* extract appears in Canto I, 'The Trystyng' (*Works*, p. 743).
Alice's comment appears in *AAW*, Ch. II, 'The Pool of Tears' (*Works*, p. 25).

Finding the Day of the Week
Carroll's method was published in *Nature* (31 March 1887) and appears in *Puzzles* 2, p. 10. We present it here for 'New Style' dates, from 14 September 1752 onwards.

Colouring Maps
This is described as a 'favourite puzzle' in *Life*, pp. 370–371. The problem was answered in 1976 by Kenneth Appel and Wolfgang Haken, who found a computer-assisted proof that four colours suffice for colouring any map; further details may be found in the author's book *Four Colours Suffice*, Allen Lane, 2002.

The Monkey and the Weight
This problem is mentioned in *Life*, pp. 317–318; *Picture Book*, pp. 267–269; and *Puzzles* 2, p. 15. Carroll believed that when the monkey starts to climb, the weight goes neither up nor down. In fact, if we ignore friction, it moves up, in such a way that the monkey and the weight remain at the same level. A discussion of the problem, including the 19 December 1893 letter to Mrs Price, appears in *Diaries* 9, pp. 113–116.

Every Triangle is Isosceles
The 'proof' appears in *Picture Book*, pp. 264–265; and *Puzzles* 2, p. 43. The flaw is in the diagram: if you draw it carefully, you'll find that the point *F* must lie *outside* the triangle, and the proof then fails.

A Symmetric Poem
This poem is described in *Universe*, pp. 19–20.

Fit the Eighth: That's Logic

The Game of Logic (referred to below as *GL*) was published on 21 February 1887, and *Symbolic Logic*, Part I (referred to below as *SL*) on 21 February 1896, both by Macmillan. Both were republished in *Mathematical Recreations of Lewis Carroll: Symbolic Logic and The Game of Logic*, Dover Publications, 1958.

Further information about Carroll's logic can be found in Amirouche Moktefi's 'Lewis Carroll's Logic', *The Handbook of the History of Logic*, Vol. 4: *British Logic in the Nineteenth Century* (ed. D.M. Gabbay and J. Woods), Elsevier, 2008, pp. 457–508; and in two articles by Francine F. Abeles, 'Lewis Carroll's Formal Logic' and 'Lewis Carroll's Visual Logic', *History and Philosophy of Logic* 26 (2005), 33–46, and 28 (2007), 1–17.

A lengthy discussion of the logical and philosophical issues in Carroll's books for children appears in Philip E.B. Jourdain's *The Philosophy of Mr. B*rtr*nd R*ss*ll*, Allen & Unwin, 1919.

The reworkings of some of Carroll's galley proofs for *Symbolic Logic*, Part II, and a description of their rediscovery, can be found in William Warren Bartley, III, *Lewis Carroll's Symbolic Logic: Part I, Elementary, 1896, Fifth Edition; Part II, Advanced, never previously published* (referred to below as *WWB*).

Introduction
The Collingwood letter appears in *Life*, p. 301, and *Letters 2* (29 December 1891), p. 879.
The Alice quotation is from *TLG*, Ch. IV, 'Tweedledum and Tweedledee' (*Works*, p. 166).

Prim Misses and Sillygisms
The diary entry is in *Diaries 1* (6 September 1855), p. 129.
The Sillygism quotation is from *SB*, Ch. XVIII, 'Queer Street, Number Forty' (*Works*, p. 388).
The cakes example and the three syllogisms appear in *GL*, pp. 21, 85, 88, 89.
The sorites examples are from *SL* (Examples 1, 35 and 40), pp. 112, 118, 119.

The Game of Logic
The 'Universe of Cakes' remark is from *GL*, p. 6.
In replacing 'All new cakes are nice' by 'Some new cakes are nice' and 'No new cakes are not-nice', Carroll assumes that some new cakes exist. Most logicians, both then and now, do not make this assumption, and this may be why Carroll's methods, though consistent, did not become better known.
The diary entry appears in *Diaries 8* (24 July 1886), p. 285.

The remark about the number of players occurs in the preface to *GL*; the 'untempting Cakes' remark appears in *GL*, p. 14; and the method for solving the cakes example is from *GL*, Ch. 1, Section 2, pp. 21–24.

The American cousins quotation is from *GL*, p. 9.

Symbolic Logic

The letter (to Mrs V. Blakemore) appears in *Letters* 2 (23 May 1887), p. 680.

The recollection from the pupil is by Evelyn Hatch, *A Selection from the Letters of Lewis Carroll to his Child-friends*, Macmillan, 1933, p. 6.

The Eastbourne child-friend was Irene Barnes; the extract from her autobiography *To Tell my Story* (1948), pp. 18–20, appears in *Diaries 8*, p. 357.

The quoted passage from *The Cambridge Review* appeared on 25 May 1887.

The letter to Mary Brown is in *Letters* 2 (21 August 1894), p. 1031.

Carroll's encouragement to his readers appears in the introduction to *SL*, pp. xvi–xvii.

The quotation is from a letter to Beatrice Griffiths (22 September 1896) in *Diaries 9*, p. 276.

The nut-cracker quotation is from the introduction to *SL*, p. xvii.

'The Problem of the School-Boys' appears in *SL*, 'Appendix, Addressed to Teachers', pp. 186–187, and is discussed in *WWB*, pp. 326–331; the conclusion is that 'None of the monitors are asleep'.

The 'Method of Trees' uses tree-diagrams (like family trees) to display all the possibilities. It reappeared in the mid-twentieth century — again, Dodgson was many years ahead of his time.

Venn, Carroll and Churchill

Logic diagrams were used even earlier than Euler, by Gottfried Wilhelm Leibniz in 1700. Dodgson discusses the differences between his own diagrams and those of Euler and Venn in *SL*, 'Appendix, Addressed to Teachers', pp. 173–183; and in *WWB*, pp. 240–249.

It is possible to draw diagrams for any number of letters, as shown by A.W.F. Edwards in *Cogwheels of the Mind: The Story of Venn Diagrams*, Johns Hopkins University Press, 2004; the Churchill example and the quotation appear there on p. 10.

The Venn quotations are from *SL*, 'Appendix, Addressed to Teachers', p. 175, and the Carroll quotation is from p. 176.

Logic Puzzles and Paradoxes

Crocodilus appears in *WWB*, p. 425, and is discussed on pp. 436–438.

The men's heights and asparagus examples are in *WWB*, pp. 427–428.

The barber's shop problem was reproduced in full as 'A Logical Paradox', *Mind*, NS, Vol. III (July 1894), pp. 436–438. It was reprinted in *WWB*, pp. 428–431, and is discussed at length on pp. 444–465.

Dodgson's diary entry appears in *Diaries 9* (31 March 1894), p. 137.

What the Tortoise Said to Achilles

'What the Tortoise Said to Achilles' was published in *Mind*, NS, Vol. IV (April 1895), pp. 278–280 (*Works*, pp. 1104–1108). There is a discussion of the paradox in *WWB*, pp. 466–470.

The Mock Turtle's 'taught-us' pun appeared in the introduction, Scene 1.

Conclusion: Math and Aftermath

Pillow-Problems (referred to below as *PP*) was published by Macmillan in July 1893, and republished in *The Mathematical Recreations of Lewis Carroll: Pillow Problems and A Tangled Tale*, Dover Publications, 1958.

Pillow-Problems

The first two quotations appear in *PP*, introduction to the first edition; the third appears in the preface to the second edition.

Carroll's answers to the selected problems are as follows; full solutions can be found in *PP*, pp. 34–35, 36–37, 39, 47, 53–54, 72.

8. 7 men; 2 shillings.

10. Either 2 florins and a sixpence, or else a half-crown and 2 shillings.

12. If s = semi-perimeter, m = area, v = volume; then
$$a^2 + b^2 + c^2 = 2(s^2 - v/s - m^2/s^2)$$

24. $DO/DA + EO/EB + FO/FC = 1$; whence any one can be found in terms of the other two.

25. $\epsilon + \alpha + \lambda - 2$.

32. (1) $n \cdot (n + 1) \cdot (2n + 13)/6$;   (2) 358 550.

49. Two.

57. The solution appears in the page of the manuscript reproduced on p. 202.

Sums of Squares

The diary entry appears in *Life*, p. 294; and *Diaries 8* (31 October 1890), p. 539.

The solution of Pillow-Problem 29 is

$$(a^2 + b^2) \times (x^2 + y^2) = (ax + by)^2 + (ay - bx)^2 = (ax - by)^2 + (ay + bx)^2.$$

Number-guessing

The diary entry is from *Diaries 9* (3 February 1896), p. 237.

The manuscript page of the number-guessing puzzle is reproduced in *Pamphlets 2*, p. 289.

The errors in Dodgson's number-guessing puzzle, and the appropriate corrections, were pointed out by Richard F. McCoart, who presents a full analysis of the puzzle in 'Lewis Carroll's Amazing Number-guessing Game', *College Mathematics Journal* 33 (November 2002), 378–383.

Divisibility

The diary entries appear in *Diaries 9* (27, 28 September, 26, 30 November and 7 December 1897), pp. 341–342, 353. Details of his methods for long division are to be found in *Picture Book*, pp. 247–263.

Right-angled Triangles

The diary entry appears in *Diaries 9* (19 December 1897), p. 354.

Epilogue

Dodgson's last diary entry appears in *Diaries 9* (23 December 1897), p. 354. Francis Paget had been Dean of Christ Church since 1892, and was later Bishop of Oxford. This quotation from his sermon appears in *Life*, p. 349.

Dodgson's mathematical self-portrait is from one of his *Four Riddles* (*Works*, p. 802).

Acknowledgements and Picture Credits

The extracts from Charles Dodgson's diaries and letters are reproduced courtesy of A.P. Watt Ltd on behalf of The Trustees of the C L Dodgson Estate and The Executives of the estate of Roger Lancelyn Green. Thanks for permission to reproduce extracts from Dodgson's letters are also due to The Alfred C. Berol Collection, Fales Library, New York University; The Houghton Library, Harvard University; and The Henry W. and Albert A. Berg Collection of English and American Literature, The New York Public Library, Astor, Lenox and Tilden Foundations.

The drawings from *Alice's Adventures in Wonderland* and *Through the Looking-Glass* were by John Tenniel, those from *The Hunting of the Snark* were by Henry Holiday, the one from *Sylvie and Bruno Concluded* was by Harry Furniss, and those for *Hiawatha's Photographing* and *A Tangled Tale* were by Arthur B. Frost.

Frontispiece: self-portrait courtesy of The Alfred C. Berol Collection, Fales Library, New York University; p. 20, drawing by Gwen Meux in F.B. Lennon's *Victoria Through the Looking-Glass*, Simon & Schuster, 1945; pp. 22, 23, 72, 73, 75, 110 (lower), 200, courtesy of the Morris L. Parrish Collection, Department of Rare Books and Special Collections, Princeton University Library; pp. 25, 74, courtesy of the National Media Museum/Science and Society Picture Library; p. 26, from *The Rectory Umbrella and Mischmasch*, Cassell & Company, Ltd, 1932; p. 28, from *The Colophon*, New Graphic Series No. 2, New York, 1939; p. 29, from Leslie P. Wenham's *The Churchyard, Richmond, North Yorkshire*, October 1976; pp. 38, 42, 54, 60, 130, 187, 202, courtesy of The Governing Body of Christ Church, Oxford; p. 41 from James Ingram's *Memorials of Oxford: The Engravings by John Le Keux*; p. 45, from *The Graphic*, 1862; p. 46, courtesy The Principal and Chapter, Pusey House, Oxford; p. 49, from Nicolas Barker's *The Oxford*

University Press and the Spread of Learning: An Illustrated History, Clarendon Press, Oxford, 1978; p. 52, from Thomas Hinde's *Lewis Carroll: Looking Glass Letters*, Collins & Brown, 1991; p. 68, courtesy of Edward Wakeling; p. 92, from C.L. Dodgson's *Euclid and his Modern Rivals*, 1879; p. 98, from C.L. Dodgson's *Curiosa Mathematica*, Part I: *A New Theory of Parallels*, 1888; p. 106, from Evelyn Hatch's *A Selection from the Letters of Lewis Carroll to his Child-friends*, 1933; p. 110 (upper), courtesy of Graham Ovenden; p. 111, from Lewis Carroll's *Alice's Adventures Under Ground*, 1886; p. 124, from *Vanity Fair*, 16 September 1876 (Liddon) and 26 February 1876 (Jowett); p. 128, courtesy of the C L Dodgson Estate; p. 142, from *The Illustrated London News*; p. 168, from Sam Loyd's *Cyclopedia of Puzzles*, Lamb Publishing Company, 1914; p. 178, from Lewis Carroll's *The Game of Logic*, 1887; pp. 186, 193, from Lewis Carroll's *Symbolic Logic*, 1896; p. 208, from Daresbury Church, Cheshire.

While every effort has been made to secure copyright, those who feel that their copyright has been infringed are invited to contact the publishers, who will endeavour to correct the situation at the earliest possible opportunity.

Index

233

About the Author

A childhood fascination with numbers led Robin Wilson to Oxford University, and to the University of Pennsylvania, where he received a PhD in number theory. He returned to his homeland to teach, first at Oxford, and then at the Open University, where he is now a professor of pure mathematics. From 2004 to 2008 he was Gresham Professor of Geometry at Gresham College, London (the oldest mathematics chair in England), and in this capacity he carried on Gresham's four-hundred-year-old tradition of making mathematics accessible to the public through open lectures. By special election he was made a Fellow of Keble College, Oxford University, and he spends one month every year running a math immersion program at Colorado College in the United States. The author and editor of over thirty books, on subjects ranging from philately to Sudoku, Robin Wilson has long been interested in exploring the human history behind the development of mathematics. When not poring over equations and theorems, Wilson can be found performing in choral societies and in the operas of Gilbert and Sullivan, and he is the coauthor of *Gilbert and Sullivan: The Official D'Oyly Carte Picture History*. He lives in Oxford.